Lecture Notes in Civil Engineering

Volume 642

Lecture Notes in Civil Engineering (LNCE) publishes the latest developments in Civil Engineering—quickly, informally and in top quality. Though original research reported in proceedings and post-proceedings represents the core of LNCE, edited volumes of exceptionally high quality and interest may also be considered for publication. Volumes published in LNCE embrace all aspects and subfields of, as well as new challenges in, Civil Engineering. Topics in the series include:

- Construction and Structural Mechanics
- Building Materials
- Concrete, Steel and Timber Structures
- Geotechnical Engineering
- Earthquake Engineering
- Coastal Engineering
- Ocean and Offshore Engineering; Ships and Floating Structures
- Hydraulics, Hydrology and Water Resources Engineering
- Environmental Engineering and Sustainability
- Structural Health and Monitoring
- Surveying and Geographical Information Systems
- Indoor Environments
- Transportation and Traffic
- Risk Analysis
- Safety and Security

To submit a proposal or request further information, please contact the appropriate Springer Editor:

- Pierpaolo Riva at pierpaolo.riva@springer.com (Europe and Americas);
- Swati Meherishi at swati.meherishi@springer.com (Asia—except China, Australia, and New Zealand);
- Wayne Hu at wayne.hu@springer.com (China).

All books in the series now indexed by Scopus and EI Compendex database!

Hemanta Hazarika · Stuart Kenneth Haigh ·
Babloo Chaudhary · Masanori Murai
Editors

Multidisciplinary Approach for Disaster Resilience and Sustainable Technologies

Springer

Editors
Hemanta Hazarika
Kyushu University
Fukuoka, Japan

Stuart Kenneth Haigh
University of Cambridge
Cambridge, UK

Babloo Chaudhary
National Institute of Technology Karnataka
Surathkal
Mangaluru, India

Masanori Murai
Shimizu Corporation
Tokyo, Japan

ISSN 2366-2557 ISSN 2366-2565 (electronic)
Lecture Notes in Civil Engineering
ISBN 978-981-96-8617-9 ISBN 978-981-96-8618-6 (eBook)
https://doi.org/10.1007/978-981-96-8618-6

This Springer imprint is published by the registered company Springer Nature Singapore Pte Ltd.
The registered company address is: 152 Beach Road, #21-01/04 Gateway East, Singapore 189721, Singapore

Contents

About the Editors

Prof. Hemanta Hazarika is currently a professor in the Graduate School of Engineering and Department of Interdisciplinary Science and Innovation, Kyushu University, Fukuoka, Japan. Professor Hazarika's research activities include disaster prevention and mitigation, soil-structure interaction, stability of soil-structures during earthquakes and tsunami, ground improvement, application of recycled waste and lightweight materials in constructions, stability of cut slopes, and landslides and their protection. He has more than 350 technical papers in various international journals, international conferences, workshops and symposia to date. He has also authored two textbooks on soil mechanics and their Japanese versions. In addition, Prof. Hazarika served as the editor of six books on diverse topics. Prof. Hazarika has several years of experience in teaching, research as well as geotechnical practice and consulting both within and outside Japan. He is currently a foreign expert of the world's first research center in Palu, Indonesia on liquefaction research called "The National Research Center for Liquefaction." Currently, Prof. Hazarika is the chairman of Asian Technical Committee on "Geotechnical Mitigation and Adaptation to Climate Change-induced Geo-disasters in Asia-Pacific Regions" of International Society of Soil Mechanics and Geotechnical Engineering (ISSMGE). He is also Founding President of a general incorporated corporation called GLOSS (Global Society for Smart Geo-Sustainnovation).

Prof. Stuart Kenneth Haigh is a professor in Geotechnical Engineering at Cambridge University and a fellow of Trinity College. In addition, Prof. Stuart Haigh is Assistant Director of the Schofield Centre. He has a wide range of research interests centered around physical modelling at 1g and utilizing centrifuges and numerical modelling using FE and MPM. His 25 years of experience in developing complex centrifuge experiments includes developing earthquake actuators, high-g robots, and an in-flight tunnel boring machine. He also delivered the Geotechnic Lecture on Mobilizable strength design for flexible retaining walls in November 2021. He has published more than 100 papers in several journals of repute.

Dr. Babloo Chaudhary is an assistant professor in the Department of Civil Engineering at National Institute of Technology Karnataka (NITK) Surathkal, India. He completed his Ph.D. from Kyushu University, Japan. He was a postdoctoral fellow at Kyoto University, Japan. His research activities include geo-disaster prevention and mitigation, dynamic soil-structures interaction, coastal geotechnics, renewal energy, ground improvement, energy geotechnics. His expertise extends from physical model tests, including 1g model tests and centrifuge model tests, to numerical simulation in the domain of geotechnical Engineering. He has published around 100 technical papers in various international and national journals, conferences, workshops and symposia. Dr. Chaudhary is recipient of Shri M. S. Jain Biennial Award given by Indian Geotechnical Society in 2017.

Dr. Masanori Murai has had a distinguished career as a senior engineer with Shimizu Corporation, a leading construction company in Japan, after receiving his Ph.D. from Kochi University in Japan. Throughout his career, he has specialized in the field of geotechnical engineering and has played a key role in the design, construction, and maintenance of a wide range of infrastructure projects. In addition to his work on infrastructure projects, Dr. Murai has also been involved in research and development activities related to geotechnical engineering. He has published numerous papers in peer-reviewed journals and presented at conferences around the world, sharing his expertise and insights with other professionals in the field. Overall, Dr. Murai's career at Shimizu Corporation has been characterized by a strong focus on safety, sustainability, and innovation. His expertise in geotechnical engineering and commitment to excellence have made him an asset to the company and the field of engineering as a whole.

The Requirements for Disaster Mitigation to Be Sustainable

Ikuo Towhata

Abstract This article presents the author's speculation on the desirable sustainable disaster mitigation for the era when the magnitude of natural disasters is getting greater. Nevertheless, disaster mitigation is not simple and still unclear because disaster mitigation is a preparedness against future problems and does not bring benefit immediately to people. It is difficult, therefore, for disaster mitigation project to be financially manageable under the current situations where many other issues, such as pandemics, environmental pollutions, support of higher educations, and many others require financial supports as well. Therefore, it is useful to combine the mitigation viewpoint with other points and make the disaster mitigation facilities be multi-purpose so that the overall income and expenditures may be in good balance. The use of inexpensive materials for which maintenance is easy is recommendable as well. Combination of public budgets with private funds is another good way so that disasters are mitigated by multiple barriers. This article picks up several examples of disaster mitigation in the past and today, and discusses their good points and lessons for future improvement.

Keywords Disaster mitigation · Sustainability · Future scope

1 Introduction

Sustainability is one of the most popular topics in a variety of expertise nowadays and the field of disaster mitigation is not an exception. The major background for this situation is that there are worldwide fears of the increasing power of nature due to the global climate change together with gigantic and devastating earthquakes that have happened since the beginning of the 21st Century, and that the people now demand more safety than in the previous century. However, most public sectors cannot afford

I. Towhata (✉)
Kanto Gakuin University, Mutsuurahigashi, Kanazawa-Ku, Yokohama, Japan
e-mail: towhata.ikuo.ikuo@gmail.com

H. Hazarika et al. (eds.), *Multidisciplinary Approach for Disaster Resilience and Sustainable Technologies*, Lecture Notes in Civil Engineering 642,
https://doi.org/10.1007/978-981-96-8618-6_1

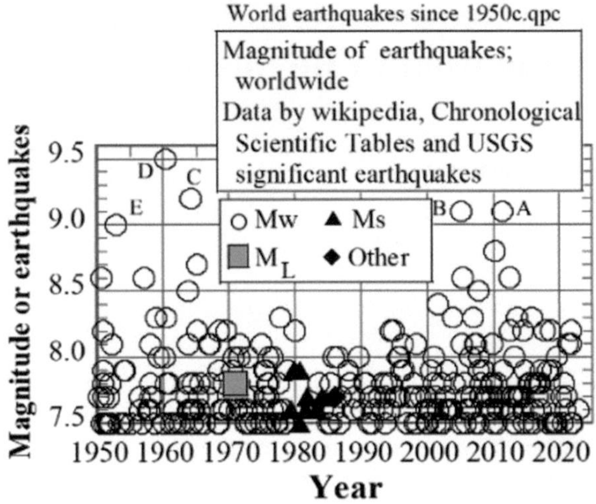

Fig. 1 History of earthquake magnitude in the world since 1950 [1]

the needed cost because of many other urgent problems such as environmental pollution, pandemics, education etc. Moreover, in spite of the hot arguments, there seems to be still insufficient view on realistic (practical) sustainable disaster mitigation. In this regard, the present paper discusses the ongoing situation and attempts to make several proposals. Prior to it, however, the present situation of natural disasters is reviewed.

Figure 1 illustrates the history of magnitudes of big earthquakes in the world. Note that the events in 1970s, 80s and 90s did not have magnitude greater than 8.3 and that this "quiet" period was the time when the major principles of earthquake engineering were developed. It is not clear whether or not those principles are good enough in the 21st Century with greater magnitudes of quakes. Human has already been faced by two devastating tsunami events in 2004 (Indian Ocean) and 2011 (Tohoku, Japan) as well as many coseismic landslide disasters in 2005 (Pakistan) and 2008 (China).

As for rainfall disasters, Fig. 2 plots the national No. 1 rainfall intensities (24-h rainfall) that were recorded during induced disasters in Japan since 1945. In line with the simple summary of rainfalls in the past, the upper bound of the intensity of disaster-triggering rainfalls has been increasing with years. This implies the increasing risk of disasters triggered by unexpected intensity of rain. Figure 3 shows the more complex nature of disasters in which the size of induced disasters in terms of the flood area used to be profound until 1970s. After 1980s, the area decreased drastically because of the human efforts to control the river water, which were construction of flood-control dams, levees, water-retention ponds, and pumping stations for drainage. In contrast, those data after mid 2010s in Fig. 3 may suggest that those human efforts started to be overcome by nature, but it is too early to make any conclusive remarks yet.

Fig. 2 History of rainfall intensity in Japan since 1945 that triggered disasters (data by JMA: Japan Meteorological Agency) [2]

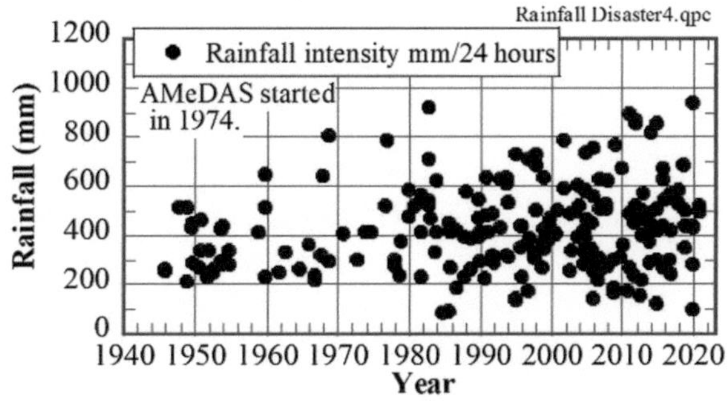

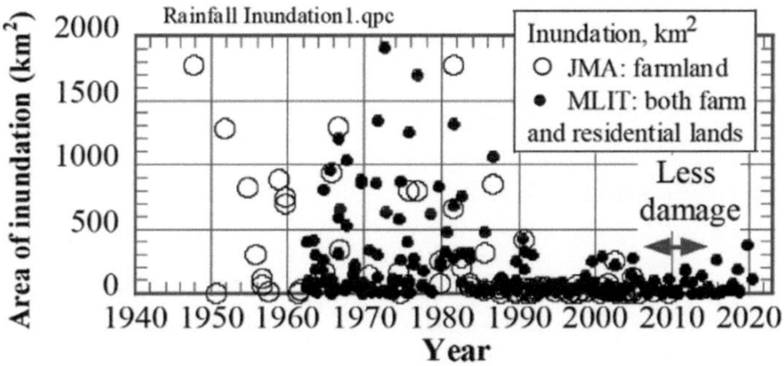

Fig. 3 Size of rainfall disasters in Japan since 1945 (MLIT: Ministry of Land, Infrastructure, Transport and Tourism) [2]

2 Value of Human Life

Because of the financial limitation within the public sectors, as mentioned before, construction of conventional types of disaster mitigation structures such as dams and high sea walls is becoming difficult in spite of the increasing power of nature. Accordingly, less expensive measures are sought for and one of the popular solutions is the disaster prediction and assistance of evacuation by using information technologies. The central concept therein is "Human life first." There is no objection to this concept but the issue is whether or not it is sufficient. In other words, consideration is needed on anything that has to be protected from natural disasters in addition to human life.

Tohni-Hongo is a fishermen's community on the Pacific Coast of Tohoku District, Japan (Fig. 4). Similar to many other communities in the same region, Tohni-Hongo was devastated by coseismic tsunami many times in the history; in 1896, 1933, 1960 and 2011. After the disaster in 1896, several families moved to higher places but they shortly came back to the original place near the coast. The reason for this was probably the inconvenience for fishing occupation as well as fire accident in the relocated place. Relocation took place in a greater scale after the 1933 event (Fig. 5).

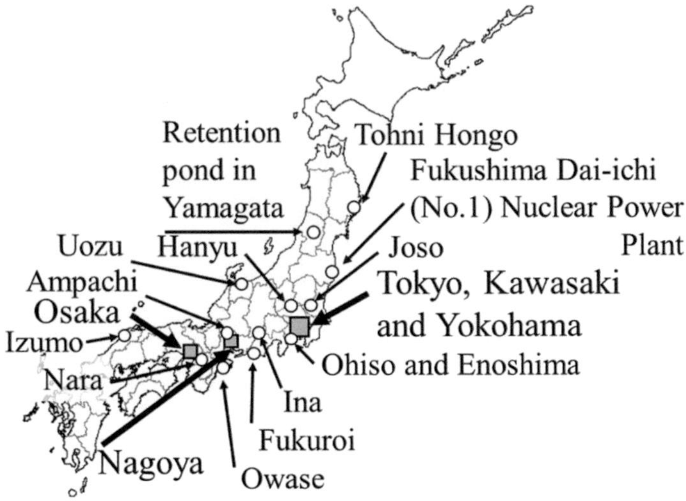

Fig. 4 Location of discussed sites in Japan

After the 1960 tsunami, the government constructed a 11.8-m-high seawall that was supposed to be higher than future tsunami. Many people felt very safe and came back to the coastal area, living in the low place behind the seawall. Pitifully, the tsunami in 2011 was higher than the wall and the overtopping water destroyed the settlement in the low land (Fig. 6). Note that an inquiry to the local people [3] suggested that those who came to the coastal area were new families who could not obtain a safe and vacant place to live in high land and were not aware of the risk of tsunami disaster in the low areas. Land price was lower near the coast. Whatever may be the reason to come to the low land, people considered the daily life and daily working (income and convenience) more important than the risk of natural hazard. This attitude of human is called normalcy bias.

The second case concerns an area along the Tama River of Kawasaki City southwest of Tokyo (Fig. 4). The confluence of a branch Hirase River with the mainstream of the Tama River is supposed to be a water retention area that is not protected by a levee. During high water in the main stream, a good amount of water enters and stays in the retention area until the flood is ceased (Fig. 7). Thereinafter, the water goes back to the main river. This flood retardation reduces the water level in the downstream area. Certainly, the area in the retention area along the Hirase River is prone to inundation as happened in 2019 and is not a good place to live. The problem is that many people are living there (Fig. 8) without fear, although the risk of inundation is demonstrated clearly to the public. The reasons for people to live in this hazardous area are;

- The present laws do not prohibit them from living or constructing condominium buildings in this hazardous area,
- Land price is relatively low,

Fig. 5 Relocated community at high place in Tohni-Hongo

Fig. 6 Tsunami devastation in 2011 behind the overtopped seawall of Tohni-Hongo

- Some people have their own shops or factories here, and
- A local railway station is very close to this area.

Similarly, in the area on the opposite side of the Tama River, a local community had declined the governmental proposal to construct a levee and suffered from inundation in 2019. The reason for decline was that a new high levee would destroy the nice river view from the restaurants and their residents. It is found here again that people

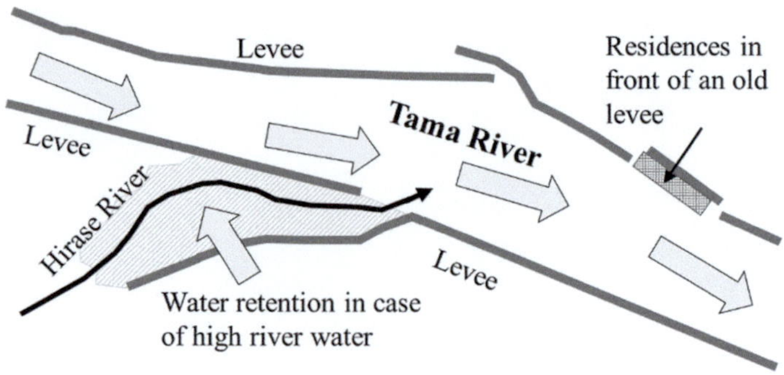

Fig. 7 Water retention area in Kawasaki along the Tama River (the water retention area is lawfully occupied by many residences, shops and factories)

Fig. 8 Residences and factories in water retention area near the Tama River

are subject to normalcy bias and put their priority to daily life and daily convenience as well as comfort rather than safety during rare disasters.

The two cases addressed above indicate that there are things other than life safety that are very important for people, which are namely daily income and daily life. People are evidently prone to normalcy bias and disaster education is not sufficient. However, it is also true that, for survival, people have to live their lives in their present living places.

The "Human life first" concept lacks in the consideration of the aftermath situation. The survived people immediately need places to live and places to work for income. Once these places are destroyed by a disaster, people cannot remain in

Fig. 9 Abandoned farm land near the destroyed nuclear power plant in Fukushima

their home community anymore and have to move to somewhere else. Thus, the community cannot be transferred to the future generation and disappears.

The present paper attempts to show that disaster mitigation has to take care not only of the lives of the present residents but also of the future generation by protecting the minimum level of infrastructures and living/working environment. Figure 9 indicates the aftermath situation near the Fukushima Dai-ichi (No.1) nuclear station (Fig. 4) where the radioactive contamination forced residents to move out of their home town, and the reconstruction of the town is still difficult. Figure 10 presents the Kobe Harbor (Fig. 4) in 1995 when the harbor facilities were destroyed by seismic liquefaction in 1995 and their reconstruction took two years during which many major customers moved to other harbors and did not come back. In both cases, the basic working environment was lost and the community underwent serious damage.

3 Cost-Sustainable Construction and Maintenance of Infrastructures for Disaster Mitigation

The previous section discussed that sustainable disaster mitigation should take care not only of human life but also infrastructures and other built environment that are essential for the continuity of local communities. In case of private sectors, the business continuity planning addresses these issues. To shed light on details, it is necessary to classify the infrastructures and other facilities into three groups,

Fig. 10 Liquefaction damage in Port Island of Kobe Harbour in 1995

which are (a) those that are essentially important during the immediate aftermath and continuity of the nation (hospitals, head quarters of urgent activities, important transportation infrastructures that are difficult to reconstruct quickly, etc.), (b) small structures that do not affect big size of communities but are still important for small localities, or can be quickly reconstructed after disasters (small bridges that have alternative bridges in their vicinity, small retaining wall in non-residential areas, etc.), and (c) others that are less important. Although this classification is not an easy task, presumably river levees and slope stabilization measures belong to Group (b). Because of its extreme importance, Group (a) can receive sufficient funding for its disaster resilience and is excluded from further discussion. There is no good reason to spend special expenditures on protection of Group (c), unless there is any special reason. Thus, Group (b) is the target of sustainable protection from disasters. Accordingly, Group (b) requires more discussion on financial aspects of their disaster management.

Group (b) consists of a huge range of infrastructures undergoing different extents of safety requirements. Figure 11 demonstrates a levee in Osaka (Fig. 4) that was damaged by the 1995 Kobe earthquake ($Mw = 6.9$). Until mid-1990s, seismic design was not considered necessary for river levees because earthquakes and heavy rains (high water level) were unlikely to occur at the same time in a probabilistic sense. Preparation of quick restoration to the original height, in case of any seismic damage, was the chief issue. This traditional idea was declined by this event in this figure because this area is close to the sea where high tide occurs twice a day, showing high probability of simultaneous occurrence of earthquake and high water. Moreover, possible overtopping of water would have inundated a vast urban area easily where the ground elevation was lower than the sea level (consequence of water pumping and consolidation settlement). Consequently, this section of the levee was reconstructed by the governmental fund with expensive ground improvement by soil–cement mixing. Thus, this case deserved full involvement of the governmental funding.

In contrast, the majority of levees cannot be retrofitted immediately in spite of the risk of climate change and heavier rainfall (Fig. 12) because of the financial restriction. In case that vulnerable local communities desire quick action for safety, efforts of self-defense is recommended.

Fig. 11 Yodo River levee in Osaka that seismically subsided due to subsoil liquefaction

Fig. 12 Site of breaching of Kinu River levee in Joso City (Fig. 4), Ibaragi, Japan (2015)

Typhoon Fran in 1976 caused more than 1000 mm of rainfall in the upstream area of the Nagara River and the levee in Ampachi Town (Fig. 4) breached as a consequence of seepage and piping (without overtopping) [4, 5]. Traditionally, this area along three major rivers, including the Nagara River, had been hit by many flood disasters. Until nineteenth century, the public sector did not seriously protect the localities from floods and the local community constructed their own levees surrounding villages. After the 19th Century, the central government took responsibility of protection from natural disasters and started to construct long levees. Since then, the length and height of the governmental levee have been improved with time. Consequently, many local communities felt safe and decided to remove their own levees because the levee inside the communities was an obstacle to traffics. Thus, at the time of the 1976 flood, the local communities relied on a single defense and the failure of the governmental levee could induce the overall flood.

There were a few smart communities that maintained their own levee (Fig. 13). This levee stopped the flood water and protected the communities from the disaster. Figure 14 is an intersection of this local levee and the main road where there is a provision to install plates to stop water flow here. Furthermore, a wealthy family had been living atop its own embankment (Fig. 15), protected by a multi-defense system that consisted of the governmental levee, a community's levee and own embankment. Thus, community and personal efforts to attain safety should be promoted by the social system.

Fig. 13 Wanouchi community's own traditional levee

Fig. 14 Gate at intersection of a community's own levee and main road

Fig. 15 Family living on private embankment

4 Construction Materials for Less Expensive Disaster Mitigation

The use of less expensive materials deserves attention in sustainable disaster mitigation. Although mixing waste materials with soil has been a common topic for improvement of liquefaction resistance of loose sand [6], the present paper introduces other materials that are more easily accepted by people.

The first material is soil which is a natural material and remain unchanged for a substantially long time. Obviously, its life time is longer than that of steel. Figure 16 shows an earth mound in Fukuroi (Fig. 4), Japan, that was constructed by a local community after its high-tide inundation and devastation in AD 1680. The advantage of this mound was its easy construction by local people and little need for maintenance (refilling of eroded portions and annual grass cutting). The present city government of Fukuroi follows this tradition and is constructing new earth mounds as tsunami shelters (Fig. 17). This type of shelter is important in this local community because there is no hill or upland in its vicinity. It is possible to use the new mound as children's playground as well. This is an example of multiple-purpose structure for disaster mitigation which will be discussed later. In addition, Fig. 18 presents the governmental seawall that is prepared against tsunami in near future. Thus, this locality is protected from tsunami disaster by a double-defense system.

Fig. 16 Historical earthen mound as a shelter during high tide (Fukuroi, Japan)

Fig. 17 Recent earthen mounds as a shelter during high tide (Fukuroi, Japan)

Fig. 18 Governmental seawall along the Pacific coast of Fukuroi (approximately 8 m high) in preparation for future tsunami disaster

Timber is another material that deserves attention. First, as is well known, the use of wood for construction contributes to reduction of carbon emission as compared with the use of steel or concrete. Second, however, durability of timber is of general concern and it is necessary to present verification of longevity of timber.

Figure 19 shows the practice of timber piles as mitigation of liquefaction risk in loose sandy ground. The idea is that loose sandy soil is densified by installation of timbers in a displacement manner (volume of the pile is equal to that of soil that is displaced laterally and compacted). Although the piles should be driven by percussion mechanism for the best efficiency, the induced noise and ground vibration make its practice difficult in urban environment. Thus, the alternative procedure starts

(a) Opening a hole by pushing soil laterally (no digging of soil)

(b) Installation of timber pile by static force

Fig. 19 Densification of loose sandy ground by installation of timber piles (Tobishima Corporation)

with opening holes in a displacement manner, followed by pile installation. Because of the silent procedure, it is possible to install timber piles in a narrow space in urban environment (Fig. 20).

The longevity of timber is achieved either by resin injection or water submergence. The former idea has made possible a wooden building that is protected from fire as well (Fig. 21). Because timber pile is the theme of this section, however, the following discussion addresses the evidence of the long life of water-submerged timber piles (Fig. 22), which is the case of liquefaction mitigation piles.

Figure 23 indicates a large-diameter timber pile "Uzu-Bashira" that consists of three component timbers (each with 1.3 m in diameter) and was excavated in the 21st Century at the site of the ancient Izumo Shrine (Fig. 4). It is assessed that this pile foundation was installed in AD 1248. Thus, the life of timber piles can be nearly one thousand years, if properly and continuously submerged in ground water.

Figure 24 demonstrates a trunk of tree that had been buried in river bed in Ina (Fig. 4), Japan, since the river was dammed by a coseismic landslide in AD 714 [7]. Recently, the river bed has been eroded by water, many ancient trees have been exposed and the age of the trees were determined by C^{14} method as AD 8th century or older [8].

Figure 25 presents a body of tree that was found in the sea off Uozu (Fig. 4), Japan. This tree was embedded in river sediment and then was inundated under the rising level of sea water. On the basis of an archaeological evidence, the time of sea water submergence is estimated to be 2500–3000 years ago [9]. Thus, if continuously

Fig. 20 Installation of timber piles in a narrow urban space (Tobishima Corporation)

Fig. 21 Wooden fireproof building in the center of Yokohama, Japan

Fig. 22 Arrow showing the removal of wooden piles of demolished structure that were encountered during construction of new buildings (Tokyo)

Fig. 23 Uzu-Bashira timber pile of ancient Izumo Shrine

Fig. 24 Trunk of a tree that was embedded in sediment of a coseismic landslide dam in or prior to the 8th century (Ina)

submerged in ground water without supply of oxygen, timber hardly decays with time. This is the reason why many buildings were constructed upon timber piles before concrete or steel piles became popular.

Apart from the issue of disaster mitigation, another advantage of timber as a sustainable construction material is indicated in the old temple buildings in Figs. 26 and 27. These temple buildings used to be a governmental building or a part of other bigger temple in the 6–8th Centuries in Nara, Japan [10]. When those original buildings became useless, they were disassembled into wooden pieces, transported to the present sites, and reassembled to the present shape. This flexibility made possible the long-time use of the timber resources. This is a good example of sustainability as compared with concrete or steel structures.

To contribute to disaster mitigation more efficiently by the use of timber, it is important to use domestic timbers and financially support the domestic forest industries. This is because well-treated forests (Fig. 28) can hold rain water, protect the slope surface from erosion, and accordingly reduce the risk of landslides particularly during heavy rain. From this viewpoint, it is not advisable to cut cheaper trees in tropical jungles in a huge scale without control, although money is saved. Therefore, in case that tropical timbers are imported, the forest's value of disaster mitigation should be converted to equivalent money and be paid to local forest owners or local timber industries.

Fig. 25 Intact shape of wood that has been buried in sea bed and submerged in water for more than 2000 years (museum exhibition in Uozu)

Fig. 26 Historical wooden building in Toshodai-ji Temple in Nara that used to be a governmental building in 8th century

5 Multi-purpose Facilities for Disaster Mitigation

To be sustainable, activities have to be financially supported and disaster mitigation is its good example. A huge retaining wall can safely prevent the collapse of a cliff but it can be installed only when it is financially feasible. When the overall budget for disaster mitigation is not sufficient, the choices are either installation of less expensive measures or finding other financial resources. The previous chapter concerned the first choice by introducing inexpensive materials, while this chapter touches upon a new financial resource. In other words, budgets for disaster mitigation and any

Fig. 27 Historical wooden building in Ganko-ji Temple in Nara; part of this building was originally constructed elsewhere at the end of 6th century [10]

Fig. 28 Well-maintained forest of cedar and cypress (Owase in Mie, Japan)

Fig. 29 Tone river levee with road at its top (Hanyu, Japan)

other purpose should be combined to install a double-purpose facility. A well-known example of this kind is a river levee with motorway at its crest (Fig. 29). Someone may point out the difficulty in merging two kinds of public budget into one project. The author believes that it is a problem of regulations made by human and that human itself can change the regulation to solve the problem.

Figure 30 shows a different kind of tsunami shelter that is called tsunami evacuation tower. The idea is that those who are enjoying the beach can escape from the tsunami wave and stay at the top of this high structure. Since it is made of steel, the sufficient mechanical resistance against tsunami pressure can be attained easily. The big spaces among steel members reduce the total tsunami force. In spite of these advantages, the problem is the need for frequent maintenance (painting) and decay of steel. The decay is particularly significant with steel structures near the sea (salty water and wind). The local community has to continuously provide fund for maintenance and renewal of the tower. This situation is not sustainable.

A more sustainable tsunami shelter is presented in Fig. 31 in which the condominium buildings along the beach are called tsunami evacuation buildings. These buildings were constructed with the owners' money and, under normal conditions, are occupied and maintained by the owners. In case of tsunami alert, people in the beach are allowed to enter the buildings and stay in the higher floors for safety. This type of shelter is called multi-purpose shelter in the present paper and is a good way to achieve safety with limited budget.

A similar tsunami facility is shown in Fig. 32 in which the building of a local government is designated as a tsunami shelter in case of alert. This building was

Fig. 30 Tsunami evacuation tower in Oh-iso, Japan

Fig. 31 Condominium building as a tsunami evacuation building (Enoshima, Japan)

Fig. 32 Local government building as a tsunami evacuation building (Kanazawa-Ku, Yokohama, Japan)

constructed and is maintained by the administrative budget and quickly changes its function in case of emergency. Thus, it is a double-purpose structure. Furthermore, Fig. 33 illustrates a vocational school in Myanmar that serves as a shelter in case of high tide.

The aforementioned retention pond of flood water can be a double-purpose facility. In Fig. 34, the retention pond is empty and used for farming in normal time. No construction of structure or house is allowed herein so that flood water is retained here when the river water level is high. After flood, the retained water goes back into the river channel. The land of a retention pond is owned by local farmers but, when the pond started operation, the central government paid a substantial money to them to make agreement that the farmers do not claim loss of products in case of inundation. Figure 35 shows another retention pond in Yokohama. This space serves as a sports facility in normal time.

Figure 36 indicates a situation of an RC building that is supporting retaining walls on its back side. It is reasonably said that this building is a part of a slope stabilizing structure and is a double-purpose structure. It was found that the wall of the basement of this building was damaged by seepage of ground water, indicating the importance of water-proofness. A similar slope stabilizing building is shown in Fig. 37 in which a condominium building is stabilizing the slope with its rigidity and weight.

Fig. 33 Vocational school in coastal area of Myanmar that becomes a shelter during high tide

Fig. 34 Water retention pond that is used for farming in normal time (Yamagata, Japan)

Fig. 35 Sports facility in Yokohama that retains water in case of high water level in the river (Yokohama)

Fig. 36 Retaining wall supported by the weight of RC building

Fig. 37 Condominium building stabilizing the underlying slope

6 Conclusion

This paper attempts to present the realistic views of sustainable disaster mitigation whose details have not been clearly understood by the human community in spite of its familiarity and importance. By reviewing the past and current situations, the following issues can make sustainable disaster mitigation a reality.

(1) The principles that were developed in the late 20th Century may not work well today because the natural disasters are more powerful than in the previous century. New principles have to be introduced.
(2) Saving of human life is absolutely important in disaster mitigation but it is not sufficient. To transfer the built environment to future generations, protection of infrastructures to a reasonable extent is important as well. If the built environment is destroyed by disasters, the existing communities may not be sustained and transferred to the next generation.
(3) Further note that people put their priority to daily life and daily income in place of life safety. Although this attitude has a bad name of normalcy bias, disaster mitigation experts should be aware of it. In other words, the built environment has to be protected from natural disasters and, during aftermath, provide people places to live and places to work.
(4) Disaster should be mitigated not only by the governmental budgets and efforts but also by the efforts of local communities and individuals. This situation is called multiple defense system.
(5) Use of inexpensive materials such as soil and timbers helps practicing sustainable disaster mitigation

(6) Multi-purpose mitigation structures can get funded for construction and maintenance from two or more resources.

References

1. Towhata I (2023) Summary of geotechnical activities in response to the 2011 Tohoku earthquake; follow-up of my TC203 Ishihara Lecture in 2019. Soil Dyn Earthquake Eng J 164:107640
2. Towhata I (2022) Views on recent rainfall-induced slope disasters and floods. In: Sustainable geo-technologies for climate change adaptation. Springer, pp 45–79
3. Kumagai M, Minami M (2021) Examining causes of dwellings in the lowlands from the upland relocation villages due to the tsunami—based on the previous researches by Yaichiro Yamaguchi and survey results. Q J Geogr 73:77–93 (in Japanese)
4. https://www.pref.gifu.lg.jp/page/6965.html
5. Takamura H, Nishiguchi T, Kinoshita T, Fukuzono T, Ohkura H (1977) Damage reconnaissance after flood of the Nagara River caused by the September 1976 Typhoon Fran. Natural disaster research report of the National Research Institute for Earth Science and Disaster Resilience 12 (in Japanese)
6. Keramatikerman M, Chegenizadeh A, Nikraz H (2017) Experimental study on effect of fly ash on liquefaction resistance of sand. Soil Dyn Earthquake Eng J 93:1–6
7. Teraoka Y, Matsushima N, Muramatsu T (2006) Buried forest in Tohyama River, historical slope disaster being a caution of future events. Publ. Iida City and Iida Art Museum (in Japanese)
8. Goto A, Muramatsu T, Teraoka Y (2008) Dates of buried trees formed by ancient landslide-dammed lake along the Toyama-river, Minami-Shinano Nagano prefecture. In: Proceedings of the 20th symposium on chronological studies at the Nagoya university center for chronological research, XIX, pp 99–102 (in Japanese)
9. Uozu Buried Forest Museum: Buried Forest (1992) (in Japanese)
10. Mitsutani, T (2001) Tree ring dating and cultural heritages. Japanese Arts. Publ Shibundo 421 (in Japanese)

Simple Prediction of Rainfall-Induced Slope Failures Based on Digital Elevation Model and Geological Information

Akihiko Wakai, Nguyen Van Thang, Go Sato, Takashi Kimura, Takatsugu Ozaki, Kensaku Matsumoto, Nanaha Kitamura, and Shota Fukushima

Abstract Every year, heavy monsoon rains cause numerous slope failures, especially in developing countries in Asia. The authors have previously developed a simple method for predicting groundwater rise and rainfall infiltration processes on natural slopes with relatively shallow depths, based on parametric studies using the finite element method assuming a semi-infinite homogeneous slope. The vertical infiltration and lateral seepage flow were numerically modeled. In the present study, an analytical method that considers the surface flow and the subsurface flow is proposed. A recent disaster in Japan is used to validate the proposed method. The proposed method has four modules: (1) infiltration process, (2) groundwater flow in vertical and lateral directions, (3) surface water flow, and (4) slope stability assessment. The proposed model allows rainfall-induced slope failures to be simply predicted for broad areas based on an infinite homogeneous slope assumption and a simplified hydrological process. The results confirm that the proposed method can simulate actual phenomena with high accuracy.

Keywords Rainfall-induced landslide · Shallow failure · Seepage analysis · Slope stability · Digital elevation model

A. Wakai (✉) · N. Van Thang · N. Kitamura · S. Fukushima
Gunma University, Kiryu, Gunma, Japan
e-mail: wakai@gunma-u.ac.jp

N. Van Thang
Thuyloi University, Dongda, Hanoi, Vietnam

G. Sato
Tokyo City University, Yokohama, Kanagawa, Japan

T. Kimura
Ehime University, Matsuyama, Ehime, Japan

T. Ozaki
R and U Resolutions Inc., Tachikawa, Tokyo, Japan

K. Matsumoto
Shizuoka Institute of Science and Technology, Fukuroi, Shizuoka, Japan

H. Hazarika et al. (eds.), *Multidisciplinary Approach for Disaster Resilience and Sustainable Technologies*, Lecture Notes in Civil Engineering 642,
https://doi.org/10.1007/978-981-96-8618-6_2

1 Introduction

The mitigation of rainfall-induced slope disasters is one of the most important issues in Asian countries. Analysis of hydrological properties and slope stability for slope failure risk assessment usually requires accurate rainfall data. In Japan, a dense nationwide network of high-performance X-band polarimetric radar stations capable of high-resolution, near-real-time observations is used to measure rainfall. Approaches for landslide risk assessment using meteorological radar data have been recently proposed, especially those that provide rapid results [1, 2]. In this paper, a simple method for predicting rainfall-induced slope failures based on a digital elevation model (DEM), geological information, and observed weather data is proposed. High-resolution radar-observed rainfall is not considered in the developed analytical model.

Numerous studies have evaluated shallow slope failures. Most rainfall-induced landslides occur at shallow ground depths. Several models have been developed to map spatial susceptibility predictions of shallow landslides [3, 4]. Available models reveal that process-based approaches are preferred for predicting the spatial and temporal occurrence of shallow landslides during precipitation events [5]. Dietrich and Montgomery [6] studied the effects of subsurface concentration and slope gradient on slope stability based on SHALSTAB steady-state shallow subsurface flow conditions. Similarly, in SINMAP (Stability INdex MAPping), shallow landslide control was studied based on grid based DEMs by Pack et al. [7]. Baum et al. [8] employed Iverson percolation method [9] for one-dimensional (1D) vertical flow in isotropic, homogeneous, saturated materials and performed an infinite slope stability analysis in TRIGRS. This method was later extended to unsaturated soils [10]. H-Slider (hill slope-scale shallow landslide-induced debris flow risk assessment), a model proposed by the Public Works Research Institute of Japan, can be used to calculate the critical steady-state rainfall required to trigger a shallow landslide. A time-varying slope stability (TiVaSS) model for shallow landslide prediction was developed by An et al. [11] based on a numerically simple solution of the three-dimensional Richards equation for subsurface flow combined with an infinite slope stability analysis. Although existing models provide good predictions of shallow landslides, using computer codes to make predictions over large areas does not provide enough spatially distributed information about the mechanical properties of such areas [5]. Generally, infiltrated rainwater moves downward through unsaturated soil layers during rainstorms, causing an increase in soil moisture content and a simultaneous decrease in soil suction. The relationship between soil saturated permeability and slope failure mechanism has been discussed by several researchers [12–14]. There is general agreement that shallow slopes with high saturated infiltration coefficients are more susceptible to failure associated with rising groundwater levels, whereas slopes with lower saturated infiltration coefficients take longer for the wet front to advance downward. Therefore, prolonged rain has a much greater effect on the latter type of slope.

Against this background, our model, named TAG_FLOW, was proposed based on a simplified hydrological approach for shallow-depth slopes [15]. It is an alternative to the finite element method for comprehensive simulations of surface and subsurface flows over shallow-depth and high-permeability slopes subjected to rainfall. The safety factor for slopes can be estimated based on the groundwater simulation results. However, the direct impact of surface water on slopes (e.g., surface erosion) is not considered in the model. In this study, the slope surface erosion caused by surface water is evaluated based on simulation results obtained using the TAG_FLOW model. A brief overview of our model and its theoretical background are first given. A validation is then performed by applying the model to a recent disaster in Japan. The results confirm that the proposed method can be used to simulate actual phenomena with high accuracy.

2 Numerical Plan of Model

In our previous study, a simplified hydrological approach for shallow-depth slopes was proposed for predicting shallow landslides using the TAG_FLOW model [15]. Based on the predicted groundwater level in slopes according to the time, the safety factor for slopes was evaluated. The assessment of slope surface erosion based on surface water flows is considered in the present study. The modeling plan is shown in Fig. 1. The surface infiltration is modeled based on the Green-Ampt model to separate the infiltration of rainwater and excess rainfall. The groundwater is predicted using a simple method developed by Wakai et al. [14] incorporated with a two-dimensional (2D) simulation of lateral groundwater exchange in the saturated soil layer. The surface water is governed by 2D shallow water equations. The infiltration and exfiltration processes control the exchange of surface water and groundwater when the groundwater level reaches the ground surface.

The safety factor for slopes is evaluated based on a simplified equation based on the infinite slope assumption. The safety factor varies according to the groundwater level. The surface flow, which promotes erosion on slope surfaces, is modeled based on the Sato-Yoshikawa-Ashida equation proposed by Sato et al. [16].

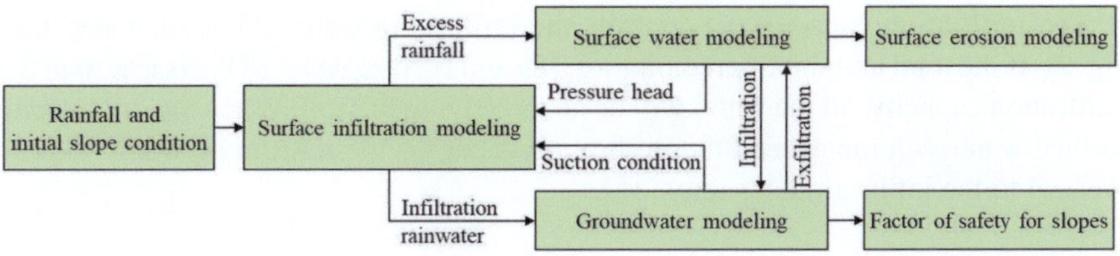

Fig. 1 Schematic diagram of modeling plan

3 Analysis and Model Development

The TAG_FLOW model for shallow landslide prediction over a wide area during rainfall was developed in Fortran 90. The model has four modules: (1) surface infiltration is modeled based on the Green-Ampt model; (2) the groundwater level is simulated based on a simple 1D vertical prediction method developed by Wakai et al. [14] combined with the horizontal groundwater distribution; (3) the surface water is controlled by 2D shallow water equations and a DEM depression removal process; (4) the stability of the slope is estimated based on the assumption of an infinite homogeneous slope [15]. A brief summary of the TAG_FLOW model components and the surface erosion module integrated into the model is given below.

3.1 Surface Infiltration Model

Surface rainfall infiltration is calculated based on the modified version of the Green-Ampt model [19] proposed by Mein and Larson [17] and Chu [18]. The equations were modified to adapt over a short time Δt of varying rainfall after the ponding condition, as shown below.

$$f(t+\Delta t) = K_s[1 + \Delta h \Delta\theta / F(t+\Delta t)] \tag{1}$$

$$F(t+\Delta t) = F(t) + K_s \Delta t + (\Delta h \Delta\theta) \ln\left(\frac{F(t+\Delta t) + \Delta h \Delta\theta}{F(t) + \Delta h \Delta\theta}\right) \tag{2}$$

$$F_p = K_s(\Delta h \Delta\theta)/(R(t) - K_s) \tag{3}$$

Here, $f(t+\Delta t)$ is the infiltration capacity at time $t+\Delta t$; $F(t)$ and $F(t+\Delta t)$ are the cumulative rainwater infiltrations at time t and $t+\Delta t$, respectively; $\Delta\theta$ is the change in volumetric water content; K_s is the saturated hydraulic conductivity; $R(t)$ is the rainfall intensity; F_p is the cumulative rainwater infiltration at the start of ponding; and $\Delta h = (h + \psi_f)$ is the driving matric pressure head, where h is the surface water depth and ψ_f is the matric suction at the wetting front.

The infiltration capacity (f) controls the infiltrating water (I) at each step (see Fig. 2). If the total amount of precipitation (R) and surface water (SW) is less than the infiltration capacity, all the water will be absorbed into the soil. When the rainfall and surface water volume exceed the infiltration capacity, the infiltrating water volume is equal to the infiltration capacity [15].

$$I = \begin{cases} \mathrm{SW} + R, & \text{if } (\mathrm{SW} + R) \le f \\ f, & \text{if } (\mathrm{SW} + R) > f \end{cases} \tag{4}$$

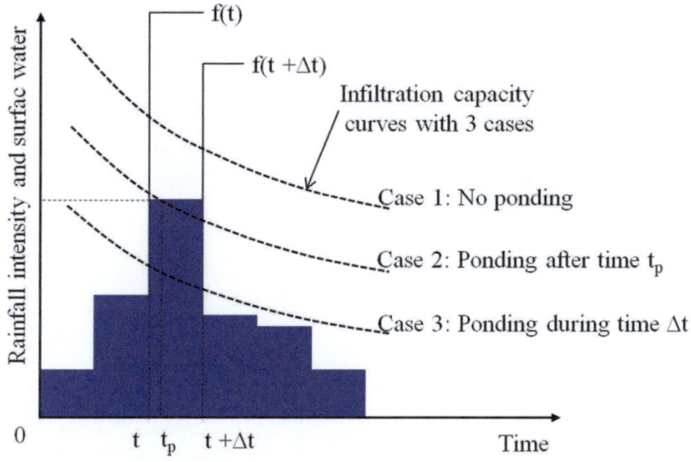

Fig. 2 Three cases in surface infiltration modeling based on Green-Ampt model

3.2 Modeling of Vertical Infiltration

The 1D vertical groundwater level is predicted based on a simple method developed by Wakai et al. [14]. In this method, the predicted groundwater level depends on the parameter θ_{cp}, which is the limit of the mean volumetric water content in the unsaturated layer before the groundwater table starts to rise (see Fig. 3):

$$\theta_{cp} = n.S_r^*/100 \tag{5}$$

$$t_1 = \frac{n(S_r^* - S_{r0})H_s}{100I} = \frac{H_s}{I}\left(\theta_{cp} - n\frac{S_{r0}}{100}\right) \tag{6}$$

$$v_{wl} = \frac{\alpha_v I}{n(1 - S_r^*/100)} \tag{7}$$

$$t_2 = \frac{H_s}{v_{wl}} = \frac{n(1 - S_r^*/100)H_s}{\alpha_v I} \tag{8}$$

where S_r^* is the critical saturation degree corresponding to the start of groundwater level rise, S_{r0} is the initial degree of saturation at the start of rainfall, n is the soil porosity, t_1 is the elapsed time before the groundwater level begins to rise, v_{wl} is the rise velocity of the groundwater level, t_2 is the elapsed time before the groundwater level reaches the ground surface, H_s is the initial thickness of the unsaturated layer, I is the infiltrating rainwater, and $\alpha_v = 2.1$ is an adjustment parameter [14].

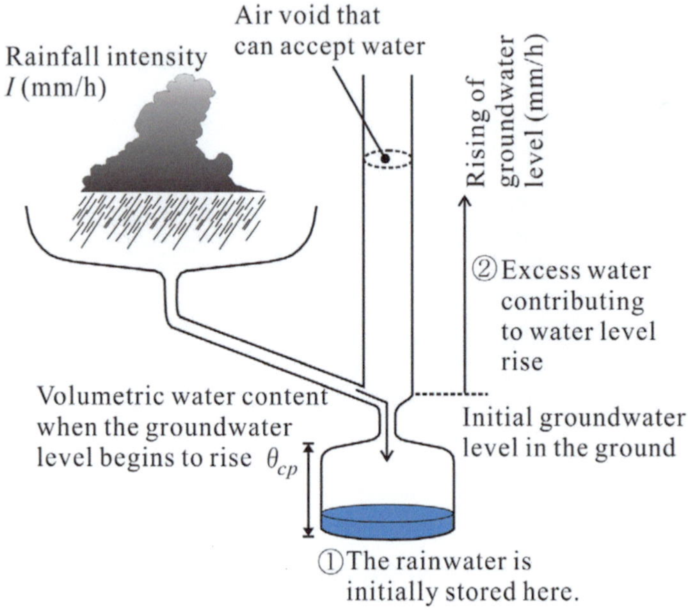

Fig. 3 Concept of developed model for vertical groundwater level rise prediction in natural slopes proposed by Wakai et al. [14]

3.3 Modeling of Lateral Seepage Flow

The groundwater redistribution is evaluated based on a governing equation in the seepage field that considers only the lateral flow in the unconfined aquifer [20]:

$$\frac{\partial}{\partial \mathrm{x}}\left(T_{xx}\frac{\partial \phi}{\partial x}\right)+\frac{\partial}{\partial \mathrm{y}}\left(T_{yy}\frac{\partial \phi}{\partial y}\right)=n_{\mathrm{e}}\frac{\partial \phi}{\partial t} \tag{9}$$

where T_{xx} and T_{yy} are the transmissivity in the x and y axis directions in the planar coordinate system, respectively, n_{e} is the effective porosity, and ϕ is the total head of the groundwater. The above equation can be easily differentiated using the explicit method proposed by Kinzelbach [21].

3.4 DEM Depression Removal Method

Surface flow modeling based on DEMs requires identifying and filling surface depressions. The identification of depressions was carried out based on the method proposed by Zhu et al. [22]. This method can be briefly summarized as follows.

(1) First, the spill elevation of the boundary cells is set equal to their original elevation.
(2) The "least cost" is the lowest value of the calculated spill elevation.
(3) Any unprocessed neighbor of the least-cost element will have its spill elevation set equal to the higher value of the least cost and its original elevation value.
(4) The next round's least-cost candidate list is updated by removing the least-cost element in step (2) and adding the neighbors processed in step (3).

A new round starts from step (2).

3.5 *Two-Dimensional Overland Flow Calculation*

The equations describing overland flow are the well-known de Saint–Venant equations [23]:

Equation of continuity:

$$\frac{\partial h}{\partial t} + \frac{\partial (hu)}{\partial x} + \frac{\partial (hv)}{\partial y} = i \tag{10}$$

Equations of motion:

$$\frac{\partial (uh)}{\partial t} + \frac{\partial \left(hu^2\right)}{\partial x} + \frac{\partial (huv)}{\partial y} + gh\left(S_{fx} + \frac{\partial H}{\partial x}\right) = 0 \tag{11}$$

$$\frac{\partial (vh)}{\partial t} + \frac{\partial (huv)}{\partial x} + \frac{\partial \left(hv^2\right)}{\partial y} + gh\left(S_{fy} + \frac{\partial H}{\partial y}\right) = 0 \tag{12}$$

where $H(x, y, t)$ is the water surface elevation above a horizontal datum, $h(x, y, t)$ is the local water depth, t is time, x, y are horizontal Cartesian coordinates, $u(x, y, t), v(x, y, t)$ are flow velocities in the x, y directions (depth-averaged flow velocities), respectively, $i(x, y, t)$ is the net distributed input rainfall, $S_{fx}(x, y, t), S_{fy}(x, y, t)$ are friction slopes in the x, y directions, respectively, and g is the acceleration due to gravity. The de Saint–Venant equations are solved using the finite difference technique as described in the study of Nguyen et al. [15].

3.6 *Slope Stability Analysis for Shallow Failures*

As shown in Fig. 4, the total safety factor (F_s) for shallow slope failures under the assumption of an infinite homogeneous slope is defined as:

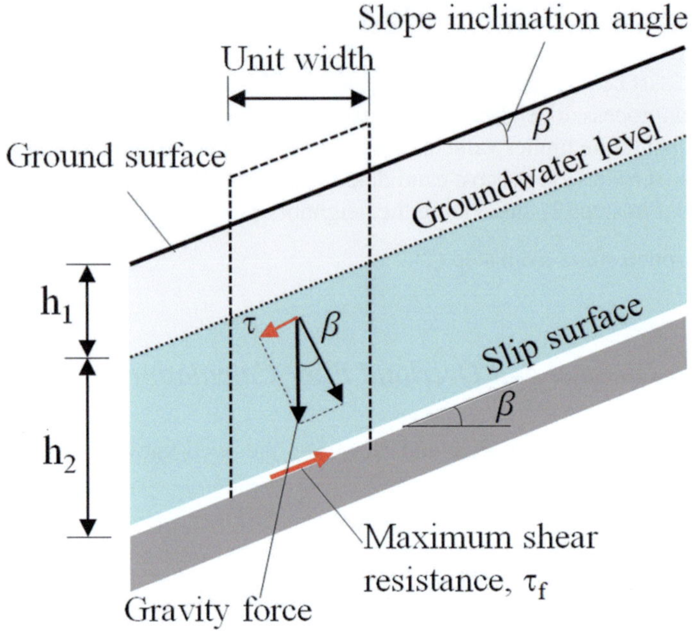

Fig. 4 Slope stability analysis for shallow failures under assumption of infinite homogeneous slope

$$F_s = \frac{\tau_f}{\tau} = \frac{c' + \{\gamma_t h_1 + (\gamma_{\text{sat}} - \gamma_{\text{w}})h_2\}\cos^2\beta\tan\phi'}{(\gamma_t h_1 + \gamma_{\text{sat}} h_2)\sin\beta\cos\beta} \quad (13)$$

where τ is the shear stress due to the sliding direction component of the gravity of soil and τ_f is the shear strength of soil, i.e., the maximum shear resistance. γ_t is the wet unit weight of soil, γ_{sat} is the saturated unit weight of soil, γ_{w} is the unit weight of water, h_1 is the depth from the ground surface to the groundwater level, h_2 is the depth from the groundwater level to the slip surface, which may correspond to the surface of the base layer, β is the slope inclination angle, c' is the cohesion of soil, and ϕ' is the angle of the shear resistance of soil.

Once the groundwater level is obtained, the stability of the slope at that time is evaluated using the value of F_s.

3.7 Evaluation of Erosion Rate

In this study, the Sato-Yoshikawa-Ashida equation proposed by Sato et al. [16] is used to calculate the slope surface erosion caused by surface water flows. This equation is based on the idea that the momentum exerted by gravity on gravel in the bed layer is equal to the momentum exerted by flow turbulence on riverbed gravel. This was theoretically investigated based on Gilbert's experimental samples at the Japan

Public Works Research Institute. The equation was proposed for the erosion volume of slope surface per unit width (q_B (cm^2/s)) as:

$$q_B = \frac{u_*^3}{(\rho_s/\rho_w - 1)g} \cdot F\left(\frac{u_*^2}{u_{*c}^2}\right) \cdot f(n) \tag{14}$$

For this equation:

$$\begin{cases} f(n) = 0.623, & \text{if } n > 0.025 \\ f(n) = 0.623 \cdot (40n)^{-3.5}, & \text{if } n \leq 0.025 \end{cases} \tag{15}$$

$$F\left(\frac{u_*^2}{u_{*c}^2}\right) = \frac{1}{1 + 8\left(u_{*c}^2/u_*^2\right)^4} \tag{16}$$

where ρ_s is the soil particle density (g/m^3), ρ_w is the density of water (g/m^3), g is the gravitational acceleration (m/s^2), u_* is the shear velocity [cm/s], u_{*c} is the critical shear velocity (cm/s), and n is Manning's roughness coefficient.

In addition, when the water flows on the river bed at a certain depth, a shear force per unit area acts on the water mass in the direction opposite to the flow. This force is balanced by the tractive force τ_0 of the water flow to the bed surface. Therefore, if the bed surface consists of non-cohesive gravel, the tractive force will pull this gravel downwards.

The tractive force per unit area τ_0 can be expressed as:

$$\tau_0 = \rho_w gRi \tag{17}$$

$$\tau_0 = \rho_w u_*^2 \tag{18}$$

Hence, the shear velocity can be written as:

$$u_* = \sqrt{gRi} \tag{19}$$

where i is the slope of the water surface, which can be approximated by the slope of the bedrock surface, and R is the radial depth, which is defined as the cross-sectional area of the water flow divided by the contact boundary between the water flow and the ground surface. In this study, since the shallow water surface is assumed to be parallel to the slope surface, the radial depth is approximately equal to the water depth. Thus, the shear velocity equation can be simplified to:

$$u_* = \sqrt{ghi} \tag{20}$$

The critical shear velocity is calculated using Iwagaki's equation [24], which depends on the soil particle size, as shown below:

$$u_{*c}^2 = \begin{cases} 0.05(\rho_s/\rho_w - 1)gd_{50}, & R_* \geq 671 \\ \{0.01505(\rho_s/\rho_w - 1)g\}^{25/22}\nu^{-3/11}d_{50}^{31/32}, & 162.7 \leq R_* \leq 671 \\ 0.034(\rho_s/\rho_w - 1)gd_{50}, & 54.2 \leq R_* \leq 162.7 \\ \left[0.1235(\rho_s/\rho_w - 1)g\right]^{25/32}\nu^{7/16}d_{50}^{11/32}, & 2.14 \leq R_* \leq 54.2 \end{cases} \quad (21)$$

where $R_* = \{(\rho_s/\rho_w - 1)g\}^{1/2}d_{50}^{3/2}/\nu$, where ν is the dynamic viscosity coefficient for water (cm^2/s) ($\nu = 0.01$ cm^2/s at 20.3 °C) and d_{50} (cm) is the corresponding particle size when the cumulative percentage reaches 50%.

The Sato-Yoshikawa-Ashida equation was proposed for calculating the erosion rate on a river bed, not on a slope surface [16]. Therefore, the effect of surface cover on overland flow needs to be considered when applying this equation to calculate soil surface erosion. In general, the surface runoff velocity decreases when the effect of vegetation cover on the slope surface is taken into account [25, 26].

Since the vegetation cover effect is not included as an input parameter in the present analysis, it is supposed that the actual flow speed is slower than the calculated one. Therefore, we controlled the amount of erosion by increasing the value of the critical shear velocity. In future, we plan to improve this in the runoff velocity calculation. The critical shear velocity was adjusted by multiplying its squared values in Eq. (21) by the dimensionless vegetation cover factor C_v.

4 Application of Model to Case Study in Murakami, Japan

4.1 Outline of Disaster

In this section, the observed slope failure distributions at the time of a recent debris flow disaster are simulated using the proposed model. It occurred in Murakami City, Niigata Prefecture, Japan, in August 2022 (Fig. 5). Due to heavy rain that started on 3 August, a huge number of slope failures and debris flows occurred in the early hours of 4 August. In the Koiwauchi district, four houses were destroyed by the debris flow; fortunately, there were no fatalities (Fig. 6). The village is located on the right bank of the Arakawa River, less than 10 km from its mouth where it empties into the Sea of Japan. Behind the village, there is a low mountain range with a ridge that is about 400 m above sea level. Most of the basement rocks of this mountain range are Neogene sedimentary rocks, with granodiorite distributed in the areas facing the river. A number of highly eroded valleys develop toward the villages in this area. A major rainfall-induced disaster, called the Uetsu Disaster, occurred around this area in 1967. The most recent disaster was much smaller than the Uetsu Disaster, but from the perspective of frequent slope failures caused by sudden torrential rain, which has become more frequent in recent years, it may have great significance in considering future disaster prevention measures in Japan. According to observations by the Automated Meteorological Data Acquisition System (AMeDAS) in Shimoseki

(Sekigawa village), about 6 km east of the affected area, the cumulative rainfall exceeded 500 mm at the time of the disaster. Around the time of the disaster, heavy rain (slightly less than 100 mm per hour) continued for several hours. This amount of rainfall is uncommon for this area.

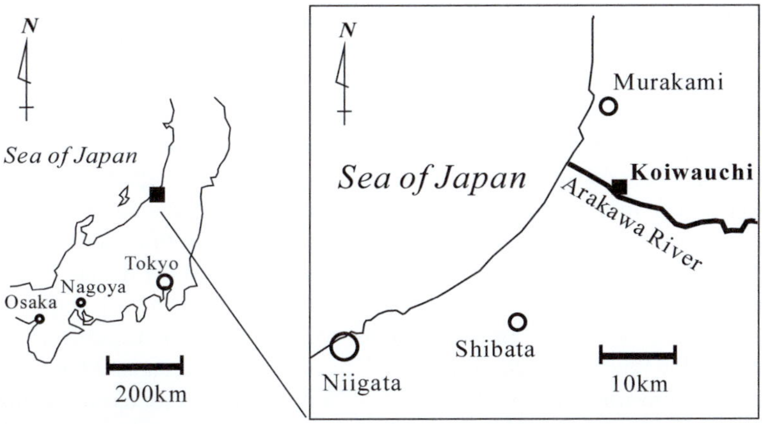

Fig. 5 Location of study area, Koiwauchi, Murakami City, Niigata Prefecture, Japan

Fig. 6 Photograph of residential land eroded by debris flow and destroyed houses in Koiwauchi village (taken on 5 August)

4.2 Study Area and Available Data

An area of about 2070 m × 1450 m that included many slope failures, shown as the white frame in Fig. 7a, around Koiwauchi district in Murakami City was analyzed using the proposed method. A topographical model was created for the analysis from 10-m meshed DEM data measured before the disaster by the Geospatial Information Authority of Japan. Figure 7b shows the distribution of the local inclination angle of the slope surface at each point for the whole analysis area. As described later, the thickness of the topsoil layer of the slope was assumed to spatially vary with local surface inclination, which can be empirically estimated. The river, the light blue region in Fig. 7b, is considered to have a flat topography and is ignored in the results shown later. The time step for the integration calculation in the analysis was set to 1 min. The local inclination angle was, as a normal function of ArcGIS, defined as the value calculated by giving the inverse square weight of the distance using the altitude value in the adjacent 3 × 3 neighborhood of the cell (Esri [27], Kamiya [28]). The rainfall data from 6:00 am on 3 August to 9:00 am on 4 August recorded by AMeDAS at the Shimoseki Observatory in Niigata Prefecture, Japan Meteorological Agency, was converted to rainfall intensity at 10-min intervals for use in the analysis. Figure 8 shows the history of the rainfall data used in the time integration. In this study, the spatial variation of rainfall intensity was not considered (i.e., the same rainfall intensity was input at all points). This variation could be considered in future studies.

According to the Seamless Digital Geological Map of Japan V2 provided by The Geological Survey of Japan, AIST [15], most of the basement rocks in the study area are Neogene sedimentary rocks, with granodiorite distributed in the areas facing the river, as shown in Fig. 9a. Table 1 shows the material parameters used in the calculations in the water flow and slope stability analysis. Since no detailed tests have been conducted in the study area, the parameters measured by Nguyen et al.

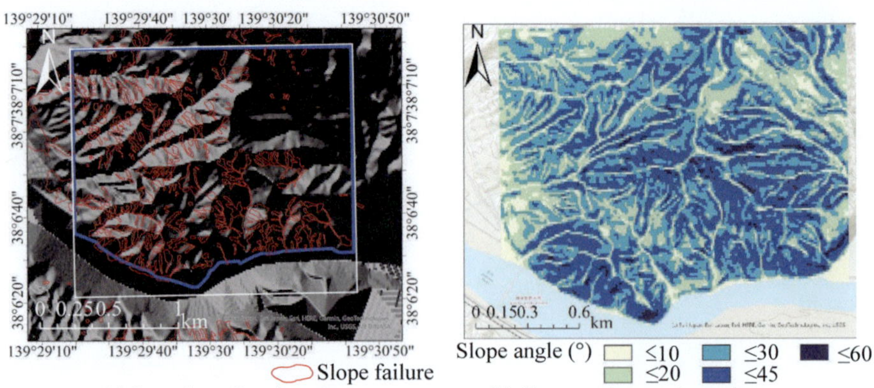

(a) Location of study area (b) Slope angle map in study area

Fig. 7 Location of study area and its topography

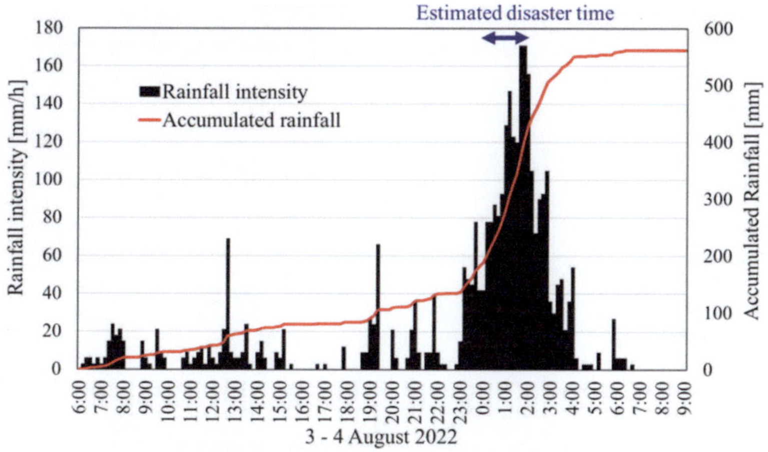

Fig. 8 Observed rainfall measured at station about 6 km east of affected area (data from AMeDAS in Shimoseki (Sekigawa village))

[15] for a soil of strongly weathered granite in Asakura City at the time of the 2017 Northern Kyushu Heavy Rain Disaster are adopted as typical soil parameters in this study. Generally, the strengthening effect of the root system of plants should be considered in the apparent shear strength of the topsoil layer; however, it was not considered in the shear strength parameters in Table 1. Tran et al. [30] stated that cohesion due to root systems plays an important role in slope stability. Wu et al. [31] estimated that the root system cohesion for Japanese pine is about 4.04 kPa and that for Korean pine is about 12.26 kPa. In this study, to consider the apparent cohesion of the topsoil layer, the minimum values of cohesion for which all the objective slopes initially exceed a safety factor of 1.0 are assumed to be given. This assumption is consistent with the fact that no slope failures occurred before the disaster. Here, the calculated value of cohesion was 8.0 kPa. This value was uniformly applied to the topsoil material.

The thickness of the topsoil layer was assumed to follow an empirical formula, proposed by Saulnier et al. [32], that depends on the slope inclination:

$$y_i = y_{\max}\left[1 - \frac{\tan(x_i) - \tan(x_{\min})}{\tan(x_{\max}) - \tan(x_{\min})}(1 - \alpha)\right] \tag{22}$$

where $\alpha = y_{\min}/y_{\max}$. This rule can be controlled by four empirical parameters, namely $y_{\min}$: smallest thickness (m), $y_{\max}$: largest thickness (m), $x_{\min}$: lowest inclination angle (°), $x_{\max}$: highest inclination angle (°). x_i is the local inclination angle defined at node i (°). In this study, following Nguyen et al. [15], $y_{\min}$ and $y_{\max}$ are assumed to be 0.2 and 5.0 m, respectively. In the analysis, assuming that the initial groundwater level is at the bottom of the topsoil layer (i.e., the top of the bedrock), the thickness of the initial unsaturated layer was set to the thickness of the topsoil

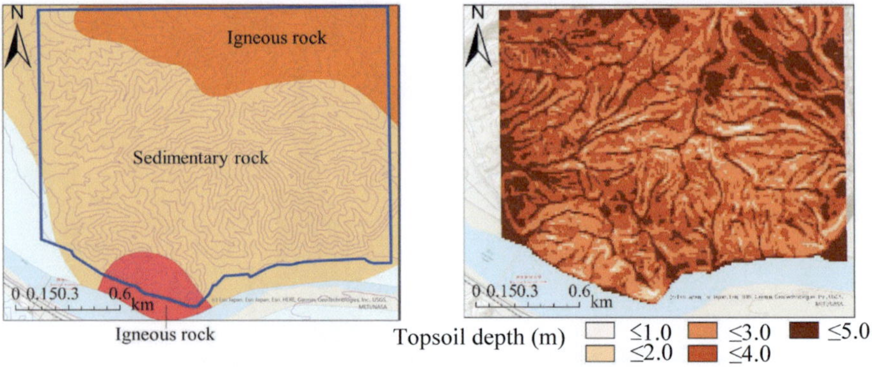

(a) Geology (b) Topsoil depth

Fig. 9 Geological conditions for study area assumed in calculations [29]

Table 1 Soil properties in study area (from Nguyen et al. [15])

Hydraulic conductivity, K_s (m h^{-1})	Saturation degree, S_{r0} (%)	Wet unit weight, γ_t (kN m^{-3})	Saturated unit weight, γ_{sat} (kN m^{-3})	Soil cohesion, c' (kN m^{-2})	Friction angle, φ' (degrees)
0.048	62.8	17.8	19.2	8.0	25.0

layer, as shown in Fig. 9b. The initial degree of saturation of the unsaturated layer was assumed to be 60% everywhere.

In this study, the original topography was processed using the DEM depression removal method to achieve stable surface water motion. The processed topography was used for the calculations. Figure 10 shows the depth of the depression estimated from the original DEM (i.e., the difference between the spill elevation for each cell and the original elevation). It can be seen that many depressions exist along the valley. Since the surface water calculations assume an initial surface water level of zero, surface water initially accumulates in the depression. After time has passed, the water level in the depression rises to the ground surface, forming a smooth channel.

The parameters used to estimate the erosion depth were set with reference to the results of soil tests reported by Nguyen et al. [15]. The values are similar to the analytical parameters used for the water flow and slope stability analysis. Regarding the parameters used in the equation for evaluating the critical shear velocity, namely Eq. (21) in Sect. 3.7, ρ_s and d_{50} are assumed to be 2.64 g/cm^3 and 0.09 cm, respectively. The critical shear velocity u_{*c}^2 calculated using Iwagaki's equation [24] was 4.918 cm^2/s^2, which may be overestimated for natural slopes with surface vegetation because this equation is based on model experiments that considered simplified phenomena where only the balance of forces acting on a single particle on a rough surface is considered. To take into account the decrease in surface water velocity due to the influence of vegetation, parametric studies with different values of the dimensionless vegetation cover factor C_v were performed. The C_v value for analytical cases

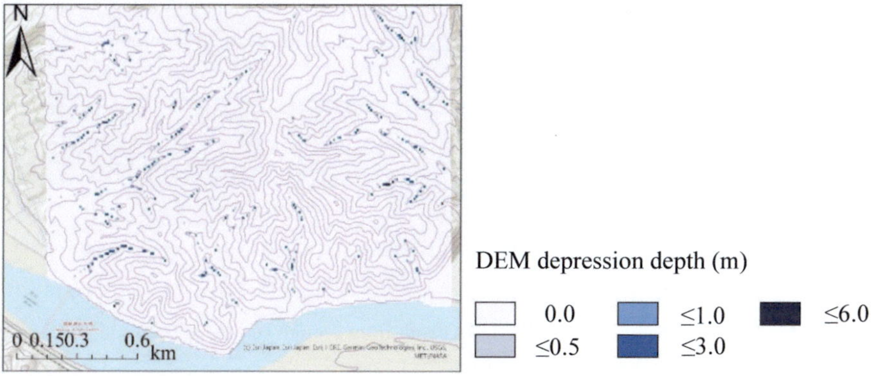

Fig. 10 DEM depression depth estimated using proposed method

Table 2 Dimensionless vegetation cover factor C_v for various analytical cases

Analytical case	Dimensionless vegetation cover factor C_v
E1	1
E2	10
E3	100

E1, E2, and E3 is shown in Table 2. For E1, u_{*c}^2 is $4.918 \mathrm{cm}^2/\mathrm{s}^2$, as estimated using Eq. (21). For E2 and E3, u_{*c}^2 is assumed to be 49.18 and $491.8\,\mathrm{cm}^2/\mathrm{s}^2$, which correspond to C_v values of 10 and 100, respectively. In the evaluation of the erosion rate, Manning's roughness coefficient n was assumed to be 0.03. $f(n)$ was set to 0.623 based on Eq. (15). The depth of surface water at depressions used in the calculations was the depth of the free water after the water level rises to the ground surface and smooth flow paths are formed.

4.3 Results and Discussion

Figure 11 shows the calculated groundwater level at 6:00 and 16:00 on 3 August and at 2:00 on 4 August. As shown, the groundwater level gradually rose as the rain continued for a long period of time. It was confirmed that during the estimated time of the disaster, the groundwater level rose with heavy rain and the stability of the slope was lost. Figure 12 shows the calculated surface water level at 0:00, 1:00, and 2:00 on 4 August. As shown, a flow path of surface water formed along the valley. In addition, the figure shows the range of sediment movement interpreted from post-disaster orthoimages and airborne laser measurement data recorded by the Tohoku branch, Japan Society of Erosion Control Engineering [33]. There are many places where the sediment movement range (red line) and the calculated flow path of the surface water well match.

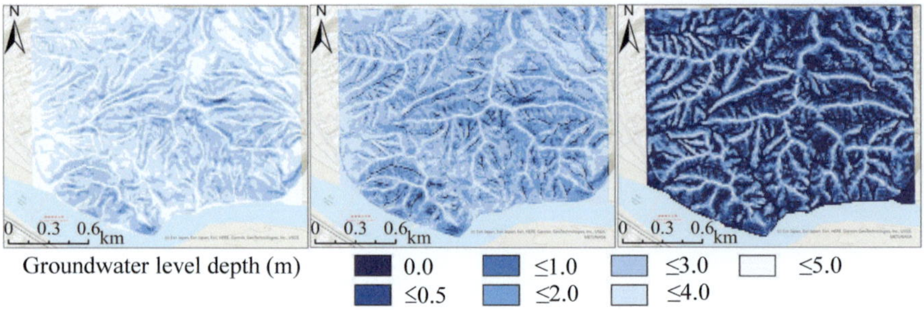

Fig. 11 Maps of calculated groundwater level depth from ground surface at 6:00 and 16:00 on 3 August, 2022, and at 2:00 on 4 August, 2022

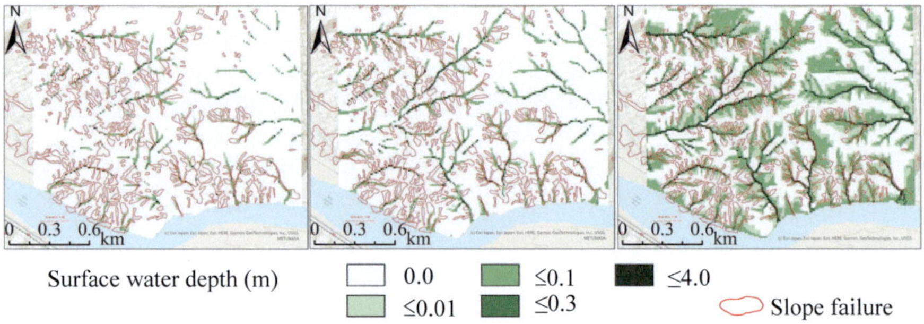

Fig. 12 Maps of calculated surface water level depth from ground surface at 0:00, 1:00, and 2:00 on 4 August, 2022

Figure 13 shows the calculated local safety factor under the assumption of infinite slope stability at 6:00 and 16:00 on 3 August and at 2:00 on 4 August. In the calculation, the parameters shown in Table 1 were set for the whole analysis area and the top of the bedrock was assumed to be a potential slip surface. This might not have reflected the actual conditions. It was confirmed that the estimated safety factor decreased as the groundwater level rose due to rain infiltration. There are many places where the sediment movement range is consistent with places where the estimated safety factor is less than 1.0. There are also many places where it is overestimated. In particular, there are places where the safety factor is less than 1.0 even though no slope failures occurred, such as the northeastern part of the analysis area. The reason for this is that although the geology is different in the northeast area shown in Fig. 13, the assumed slope conditions were applied to the whole analysis area using the same method in the analysis. There was thus a slight problem with setting the conditions.

Figure 14 shows the calculated erosion depth at 2:00 on 4 August obtained using three different critical shear velocities. While the sediment movement area and the eroded area coincided in many places, large erosion was seen even in areas where slope failures did not occur. Figure 15 shows the calculated time history of the

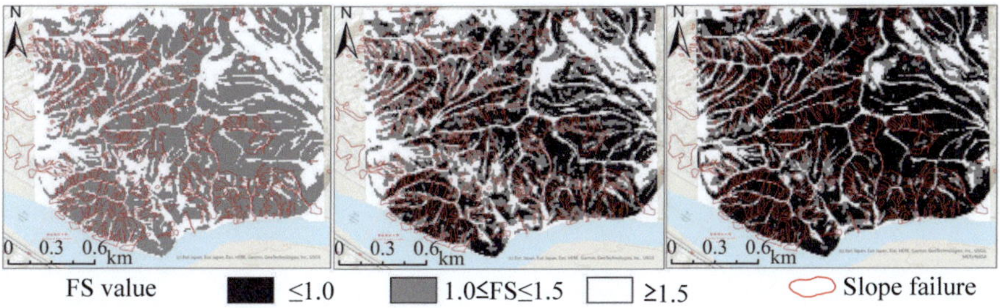

Fig. 13 Calculated safety factor maps for slopes at 6:00 and 16:00 on 3 August, 2022, and at 2:00 on 4 August, 2022

erosion amount at point A shown in Fig. 14. Since E1 is a setting in which there is no vegetation on the ground surface (it consists only of exposed soil), the calculated erosion depth tends to be large overall. Although u_{*c}^2 for E2 was 10 times larger than that for E1, the calculated erosion depth for E2 was similar to that for E1. u_{*c}^2 for E3 was 100 times larger than that for E1 and the erosion depth for E3 was smaller than that for E1 and E2. When surface water flows on natural slopes with vegetation, erosion is considered to be suppressed by the vegetation. The overestimation of erosion could be reduced by considering the influence of such vegetation. However, since no comparison or discussion of actual phenomena has been carried out, it is necessary to reexamine the critical shear velocity and try other models in future to find one that better reproduces actual phenomena.

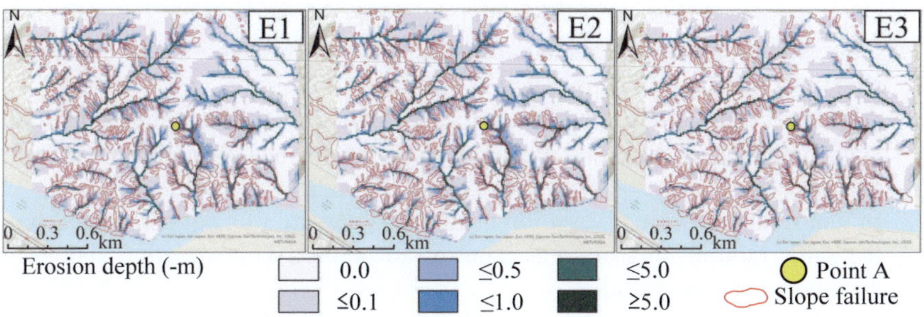

Fig. 14 Calculated erosion depth maps for E1, E2, and E3

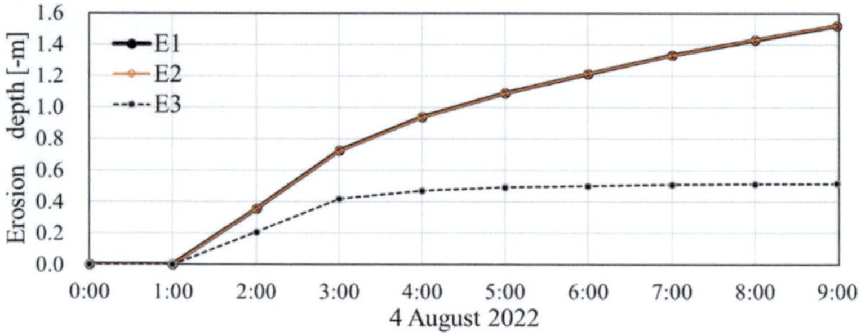

Fig. 15 Calculated time history of erosion depth at point A (see Fig. 14) for E1, E2, and E3

5 Conclusion

The primary results of the present study are as follows:

(1) The authors previously developed a simple method for predicting groundwater rise and the rainfall seepage process in natural slopes at a relatively shallow depth based on parametric studies conducted using the finite element method under the assumption of a semi-infinite homogeneous slope. The vertical infiltration and lateral seepage flow were numerically modeled. In the present study, an analytical method that considers the surface flow and the subsurface flow was proposed.

(2) The proposed method had four modules: (i) infiltration process, (ii) groundwater flow in vertical and lateral directions, (iii) surface water flow, and (iv) slope stability assessment. The proposed model allowed rainfall-induced slope failures to be simply predicted for broad areas based on an infinite homogeneous slope assumption and a simplified hydrological process.

(3) A recent disaster in Japan was used to validate the proposed method. The results confirmed that the proposed method can simulate actual phenomena with high accuracy.

(4) In this study, the spatial variation of rainfall intensity was not considered (i.e., the same rainfall intensity was input at all points). This variation should be considered in future studies.

(5) In future, we also plan to improve this in the runoff velocity calculation. When surface water flows on natural slopes with vegetation, erosion is considered to be suppressed by the vegetation. The overestimation of erosion could be reduced by considering the influence of such vegetation. It is necessary to reexamine the critical shear velocity and try other models in the future to find one that better reproduces actual phenomena.

Acknowledgements Part of this work was supported by the Japan Science and Technology Agency SICORP (e-ASIA Joint Research Program) under grant number JPMJSC18E3, titled as "Establishment of a Landslide Monitoring and Prediction System".

References

1. Okimura T, Torii N, Osaki Y, Nambu M, Haraguchi K (2011) Construction of real-time type hazard system to predict landslides caused by heavy rainfalls. J Japan Soc Erosion Control Eng 63(6):4–12 (in Japanese)
2. Kinoshita A, Kanno T, Okamoto A, Hitokoto M, Onodera M, Sakuraba M, Sugiyama M (2013) Construction of real-time hazard map system in Rokko mountain area. J Japan Soc Erosion Control Eng 66(1):15–22 (in Japanese)
3. Kim MS, Onda Y, Kim JK, Kim SW (2015) Effect of topography and soil parameterisation representing soil thicknesses on shallow landslide modeling. Quat Int 384(Mar):91–106. https://doi.org/10.1016/j.quaint.2015.03.057
4. Rosso R, Rulli MC, Vannucchi G (2006) A physically based model for the hydrologic control on shallow landsliding. Water Resour Res 42(6):6. https://doi.org/10.1029/2005WR004369
5. Raia S, Alvioli M, Rossi M, Baum RL, Godt JW, Guzzetti F (2014) Improving predictive power of physically based rainfall-induced shallow landslide models: a probabilistic approach. Geosci Model Dev 7(14):495–514. https://doi.org/10.5194/gmd-7-495-2014
6. Dietrich WE, Montgomery DR (1998) SHALSTAB: a digital terrain model for mapping shallow landslide potential. National Council of the Paper Industry for Air and Stream Improvement, Technical Rep. Corvallis, OR
7. Pack RT, Tarboton DG, Goodwin CN (1998) Terrain stability mapping with SINMAP, technical description and users guide for version 1.00. Terratech Consulting, Salmon Arm, BC
8. Baum RL, Savage WZ, Godt JW (2002) TRIGRS-A Fortran program for transient rainfall infiltration and grid-based regional slope stability analysis. Denver Publishing Service Center, Denver
9. Iverson RM (2000) Landslide triggering by rain infiltration. Water Resour Res 36(Jan):1897–1910. https://doi.org/10.1029/2000WR900090
10. Baum RL, Savage WZ, Godt JW (2008) TRIGRS-A Fortran program for transient rainfall infiltration and grid-based regional slope stability analysis, version 2.0. Denver Publishing Service Center, Denver
11. An H, Tran TV, Lee GH, Kim Y, Kim M, Noh S, Noh J (2016) Development of time-variant landslide-prediction software considering three-dimensional subsurface unsaturated flow. Environ Modell Software 85(12):172–183. https://doi.org/10.1016/j.envsoft.2016.08.009
12. Ozaki T, Wakai A, Watanabe A, Cai F, Sato G, Kimura T (2021) A simplified model for the infiltration of rainwater in natural slope consisting of fine sands. J Japan Landslide Soc 58(2):57–64. https://doi.org/10.3313/jls.58.57
13. Rahardjo H, Ong TH, Rezaur RB, Leong EC (2007) Factors controlling instability of homo geneous soil slopes under rainfall. J Geotech Geoenviron Eng 133(12):1532–1543. https://doi.org/10.1061/(ASCE)1090-0241(2007)13
14. Wakai A, Hori K, Watanabe A, Cai F, Fukazu H, Goto S, Kimura T (2019) A simple prediction model for shallow groundwater level rise in natural slopes based on finite element solutions. J Japan Landslide Soc 56(Jan):227–239. https://doi.org/10.3313/jls.56.227 (in Japanese)
15. Nguyen VT, Wakai A, Sato G, Viet TT, Kitamura N (2022) Simple method for shallow landslide prediction based on wide-area terrain analysis incorporated with surface and subsurface flows. Nat Hazards Rev 23(4):04022028. https://doi.org/10.1061/(ASCE)NH.1527-6996.0000578
16. Sato S, Yoshikawa H, Ashida K (1957) Study on bed load transport of river bed gravel. Proc Hydraul Res Soc 2:25–26. https://doi.org/10.11532/prohe1956.2.25
17. Mein RG, Larson CL (1973) Modeling infiltration during a steady rain, Water Resour Res 9(2):384–394. https://doi.org/10.1029/WR009i002p00384
18. Chu ST (1978) Infiltration during an unsteady rain. Water Resour Res 14(3):461–466. https://doi.org/10.1029/WR014i003p00461
19. Green WH, Ampt G (1911) Studies on soil physics. J Agric Sci 4(1):1–24
20. Japanese Association of Groundwater Hydrology (2010) Simulation of groundwater flow and solute transport. Gihodo Shuppan

21. Kinzelbach W (1986) Groundwater modeling: an introduction with sample programs in BASIC. Elsevier Science Ltd 1986:29–35
22. Zhu D, Ren Q, Xuan Y, Chen Y, Cluckie ID (2013) An effective depression filling algorithm for DEM-based 2-D surface flow modeling. Hydrol Earth Syst Sci 17:495–505. https://doi.org/10.5194/hess-17-495-2013
23. Di Giammarco P, Todini E, Lamberti P (1996) A conservative finite element approach to overland flow: the control volume finite element formulation. J Hydrol 175(1–4):267–291. https://doi.org/10.1016/S0022-1694(96)80014-X
24. Toichiro T (1974): Hydraulics II (Chapter 14). In: Morikita publishing basic civil engineering comprehensive book 7 (in Japanese)
25. Kitahara H (2002) Vegetation surface erosion prevention function. J Sabo Soc 54(5):92–101 (in Japanese)
26. Li C, Pan C (2018) The relative importance of different grass components in controlling runoff and erosion on a hillslope under simulated rainfall. J Hydrol 558(2018):90–103. https://doi.org/10.1016/j.jhydrol.2018.01.007
27. Esri. Raster surface toolset concepts, how slope works. https://desktop.arcgis.com/ja/arcmap/10.3/tools/3d-analyst-toolbox/how-slope-works.htm. Accessed 21 Sept 2022
28. Kamiya I, Kuroki T, Tanaka K (2000) Interpretation of geomorphology and geology map. Geoinformatics 11(1):11–24 (in Japanese)
29. Geological Survey of Japan, AIST (2022) Seamless digital geological map of Japan V2 1: 200,000. https://gbank.gsj.jp/seamless. Accessed September 12 Sept 2022
30. Tran TV, Lee GH, Kim MS (2016) Shallow landslide assessment considering the influence of vegetation cover. J Korean GeoEnviron Soc 17(4):17–31. https://doi.org/10.14481/jkges.2016.17.4.17
31. Wu TH, McKinnell WP, Swanston DN (1979) Strength of tree roots and landslides on Prince of Wales Island, Alaska. Can Geotech J 16(1):19–33. https://doi.org/10.1139/t79-003
32. Saulnier GM, Bevenb K, Obleda C (1997) Including spatially variable effective soil depths in TOPMODEL. J Hydrol 202(1–4):158–172. https://doi.org/10.1016/S0022-1694(97)00059-0
33. Tohoku branch of Japan society of erosion control engineering, "Distribution of the extent of sediment transport in Shimoi Iwanai area, Murakami City, Niigata Prefecture". https://jsece.or.jp/branch/tohoku/2022/08/12/%e6%96%b0%e6%bd%9f%e7%9c%8c%e6%9d%91%e4%b8%8a%e5%b8%82%e4%b8%8b%e5%b2%a9%e5%86%85%e5%9c%b0%e5%8c%ba%e3%81%8a%e3%81%91%e3%82%8b%e5%9c%9f%e7%a0%82%e7%a7%bb%e5%8b%95%e7%af%84%e5%9b%b2%e5%88%86%e5%b8%83/. Accessed 15 Aug 2022

Impact of Climate Change on Landslide Susceptibility and Resilience in Central Taiwan

K. J. Shou, V. C. Lai, and W. J. Lin

Abstract This study aims to investigate the landslide susceptibility and resilience of the study area located in Central Taiwan, including the mountain highway and the villages in this area. The landslide susceptibility values were calculated by Logistic Regression model. By considering the core concept of resilience, the control factors were adopted and the data was collected for the study area. In order to explore the temporal behavior of landslide resilience, the population density, electricity consumption per household, and the cost of reconstruction were adopted as the indicators to establish the landslide disaster resilience model. Based on the data layers, the comprehensive linear regression model was established and used for the resilience interpretations, with suggestions on the village human mobility and the rehabilitation budget for resilience enhancement. This study can help for landslide disaster mitigation, especially for the enhancement of landslide resilience of the villages in mountain areas.

Keywords Landslide susceptibility · Logistic regression · Resilience · Linear regression

1 Introduction

Following the Hyogo Framework for Action, initiated on the 2005 World Conference on Disaster Reduction (WCDR), Sendai Framework for Disaster Risk Reduction (SFDRR) was presented in 2015 [1], in which the disaster resilience was more clearly defined to improve the capability of recovering from a disaster, by the countermeasures in the aspects of social, economy, policy, etc. The most popular definition of resilience could be "the capability of an ecosystem enduring external disturbance and returning to the state before the disturbance". But it is still scarce for the research on the quantitative disaster resilience, including the geohazard-related resilience. In

K. J. Shou (✉) · V. C. Lai · W. J. Lin
Department of Civil Engineering, National Chung-Hsing University, Taichung, Taiwan
e-mail: kjshou@dragon.nchu.edu.tw

H. Hazarika et al. (eds.), *Multidisciplinary Approach for Disaster Resilience and Sustainable Technologies*, Lecture Notes in Civil Engineering 642,
https://doi.org/10.1007/978-981-96-8618-6_4

view of the rainfall induced landslide hazards in Taiwan, quantifying and enhancing the resilience are critical to protect the life, property, and infrastructure in the landslide area. This study comprises two major part, i.e., calculating the landslide susceptibility as a hazard factor, and collecting data of the control factors, then establishing the landslide resilience model of the research area.

2 Methodologies

The study area is located in the upstream of Wu River watershed in Central Taiwan, with focuses on the mountain highway and the six villages (see Fig. 1). Data of the landslide causative factors and the resilience control factors before and after rainfall events in study area were collected, then the landslide susceptibility values were calculated and the landslide disaster resilience models were established.

2.1 *Landslide Susceptibility Analysis*

This study adopted the Logistic Regression method of Lin [2], with the nine causative factors, i.e., elevation, slope, slope, dip slope index, distance to fault, distance to road, distance to river, greenness index, and 72 h cumulative rainfall. Among them, the hourly rainfall data during 2012–2017 was further analyzed to find the largest 72-h cumulative in a quarterly basis for each rainfall station, such that we can explore the spatial changing trend of rainfall in the study area.

Considering the villages as the analysis units, the Thiessen's Polygon method was used to obtain the rainfall distribution, as shown in Fig. 2. By introducing the seasonal rainfall data to the following Logistic Regression model of Lin [2], as described in Eq. (1), we can obtain the seasonal landslide susceptibility values of each village.

$$\begin{aligned}\ln\left(\frac{P}{1-P}\right) &= -0.577 * F1 + 1.545 * F2 + 0.077 * F3 - 0.038 * F4 \\ &\quad - 0.242 * F5 - 0.102 * F6 - 0.458 * F7 - 3.564 * F8 \\ &\quad + 0.560 * F9 - 0.406 \end{aligned} \tag{1}$$

where p is the probability of landslide. Factors $F1$–$F3$ are distance to road, distance to river, and distance to fault; factors $F4$–$F6$ are elevation, slope angle, and aspect. $F7$ is the dip slope index, $F8$ is the Greenness Index, and $F9$ is the maximum 72 h accumulative rainfall.

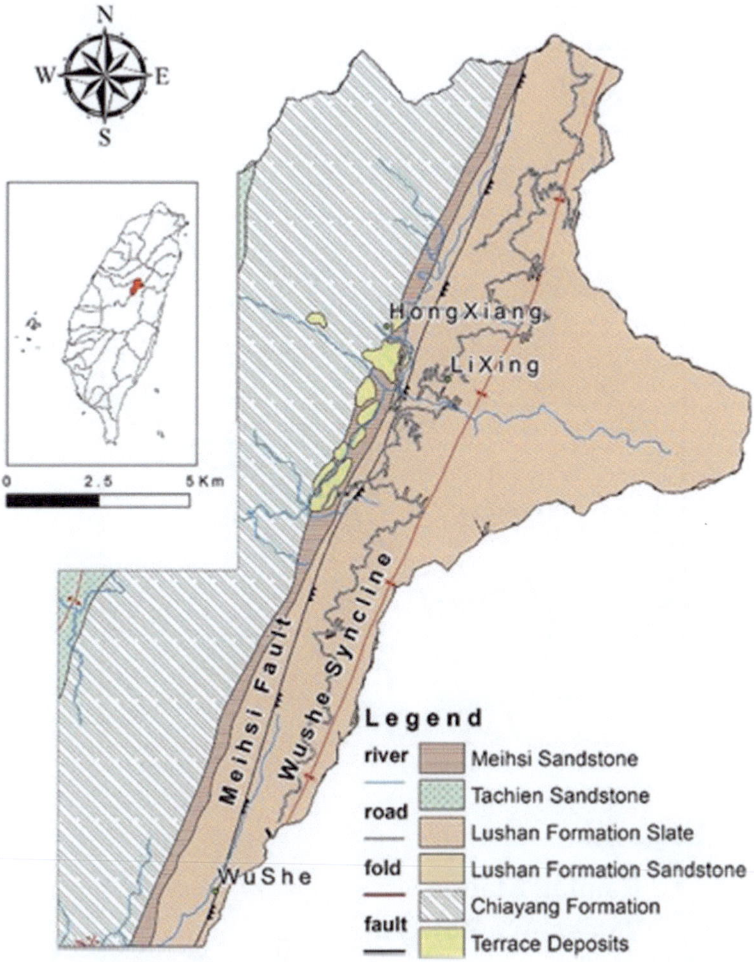

Fig. 1 The location and geology of the study area study are in the upstream of Wu River watershed in Central Taiwan

2.2 *Landslide Disaster Resilience*

Based on a review of the behavior of landslide hazard, nine control factors were selected for assessing the landslide resilience in the study, including population density, average electricity consumption per household, elder ratio, shelter ratio, fire fighter ratio, medical manpower ratio, income tax, and reconstruction budget.

This study further applied linear regression analysis based on the resilience of landslide disasters defined by Eqs. (2) and (3). Karen [3] quantified the resilience of landslide disasters with time dependence, using Eq. (2) to define the incremental resilience $r(t)$, as the cost per unit time $C(t)$ multiplying unit time $\mathrm{Inc}(t)$ and dividing

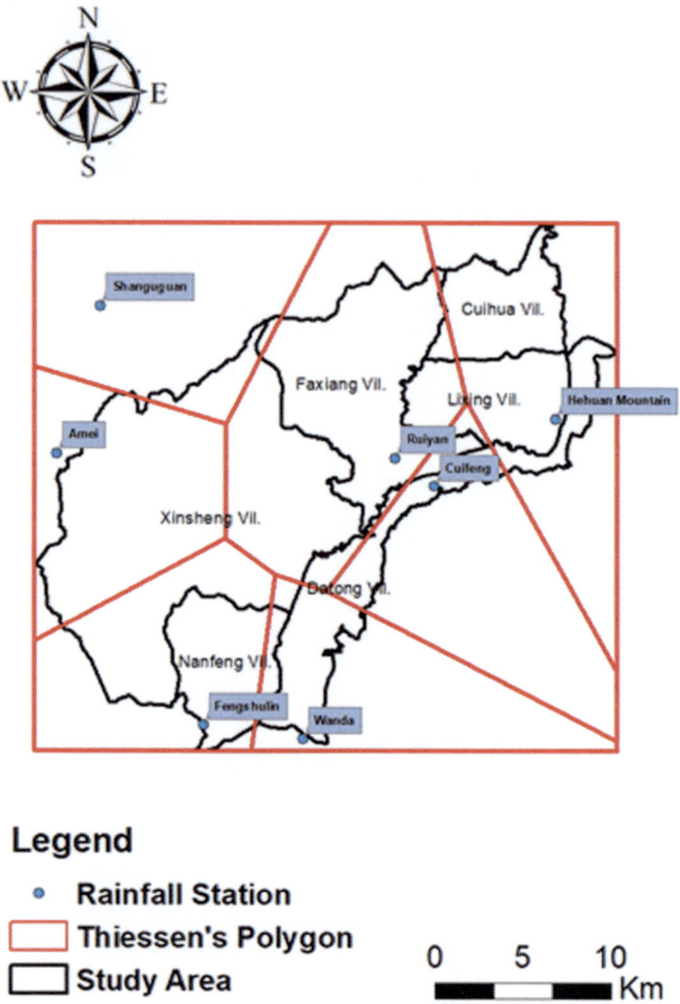

Fig. 2 The Thiessen's polygons to obtain the rainfall distribution of the six villages in the study area

by total time (NI). In Eq. (3), by integrating $r(t)$ according to time, the resilience Re can be obtained.

$$r(t) = \frac{C(t)\mathrm{Inc}(t)}{\mathrm{NI}(t)} \tag{2}$$

$$\mathrm{Re}(t_s) = \int_{t_0}^{t_s} r(t)\mathrm{d}t \tag{3}$$

3 Linear Regression Analysis

In order to more properly establish the landslide resilience model, this study tried to use different control factors and divided the data into high season and low season for linear regression analysis. Before further establishment of resilience model, the correlation of those control factors for the major resilience indicators were analyzed.

3.1 Population Density

The linear regression analysis was carried out for the high season and low season of different periods of years, namely 2012–2016, 2016–2017, 2012–2017, 2012–2015, and 2015–2017. The independent control factors include landslide susceptibility (RI1), elder ratio (RI2), shelter ratio (RI3), fire fighter ratio (RI4), medical manpower ratio (RI5), income tax (RI6), and rehabilitation budget (RI7).

The difference between low season and high season is that the impact of landslide susceptibility on population density is reduced, that is, it shows that the impact of landslides on the study area is relatively lower during the period when the population density is lower. The results suggest that, whether in high season or low season, the best model for forecasting as shown in Eq. (4) is the 2012–2015 period, as they can properly to predict the population density in 2016–2017. Therefore, the population density could be considered as a dependent factor to assess the recovering from a landslide disaster.

$$Y = -0.731 * \mathrm{RI1} - 0.594 * \mathrm{RI2} + 0.434 * \mathrm{RI3} - 0.260 * \mathrm{RI4} + 0.820 * \mathrm{RI5} - 0.543 * \mathrm{RI6} - 0.124 * \mathrm{RI7} + 1.879 \quad (4)$$

3.2 Electricity Consumption

Similarly, the linear regression analysis for the resilience indicator, i.e., electricity consumption, was carried out for the high season and low season of different time periods. Comparing the R^2 coefficients of five models of electricity consumption, i.e., 0.597–0.436, it shows the regression is barely acceptable. The weightings of the control factors on the electricity consumption for the 2016–2017 model as shown in Eq. (5) can be ranked as: landslide susceptibility, medical manpower ratio, shelter ratio, reconstruction budget, income tax, elder ratio, and fire fighter ratio.

$$Y = -0.199 * \mathrm{RI1} - 0.617 * \mathrm{RI2} + 0.419 * \mathrm{RI3} - 1.310 * \mathrm{RI4} + 0.820 * \mathrm{RI5} - 0.505 * \mathrm{RI6} - 0.138 * \mathrm{RI7} + 1.879 \quad (5)$$

Table 1 The prediction errors of regression models for different seasons in 2016–2017

Year	Season	% of error				
		Xinsheng	Lixing	Faxiang	Nanfeng	Cuihua
2016	1	2	3	10	45	8
	2	6	5	9	46	7
	3	6	5	12	44	17
	4	4	6	18	53	12
AVE		5	5	12	47	11
2017	1	6	19	8	8	28
	2	2	19	1	24	19
	3	3	27	8	10	23
	4	1	26	2	13	27
AVE		3	23	5	14	24

Based on the highest R^2 value, it suggested the best mode for both high season and low season is the 2012–2015 model. This model was applied to predict the population density of 2016–2017 with error between 1 and 53%, best for Xinshen village and worst for Nanfeng village. For the low season, the error is between 0 and 66%, 0% for Faxiang village and 66% for Nanfeng village. The results suggest there are variation of error temporally and spatially (Table 1).

4 Lanslide Resilience Analysis

This study applied the model of Karen [3] to study the resilience of landslide disasters, to explore the differences in reconstruction budget in different time scales.

4.1 Model Based on Reconstruction Budget

In order to compare the impact of reconstruction budget on the resilience, the linear regression analysis was carried out by letting reconstruction budget as the dependent variable. The cases of analyses were numbered (see Table 2). The linear regression analysis model with R^2 of 0.623 was used to estimate the 2012–2017 reconstruction budget (see Fig. 3); then introduce the data into the formula of Karen [3] to obtain the resilience, as shown in Figs. 4, 5 and 6. The results show that the resilience of Lixing, Datong, Nanfeng and Cuihua villages changed from negative to positive in the post-2015 period, while that of the remaining two villages (Xinshen and Faxiang) was positive and increasing. In terms of the concept of disaster resilience, the lower the cost of rehabilitation, the higher the resilience, indicating that its villages do not

Table 2 Different group settings for the resilience analyses

Case	Groups of data	
A	2008–2011	2012–2016
B/C	2012–2015	2016–2017

need too much money to deal with natural disasters. However, it is suggested to have more evidence to support the findings of this study.

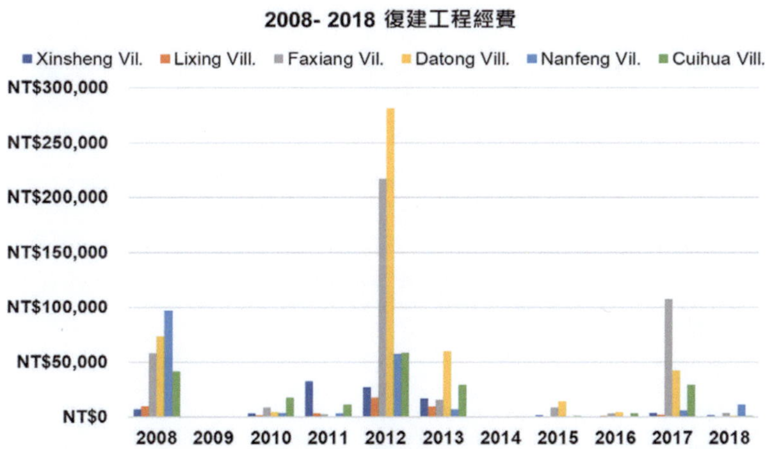

Fig. 3 The 2008–2018 reconstruction budgets for the villages in the study area

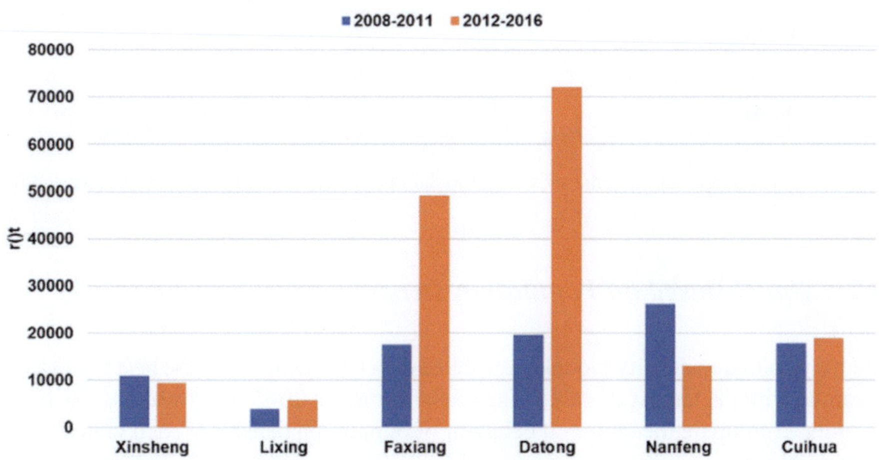

Fig. 4 The resilience values of the six villages in the study area, for Case A with the reconstruction budget model

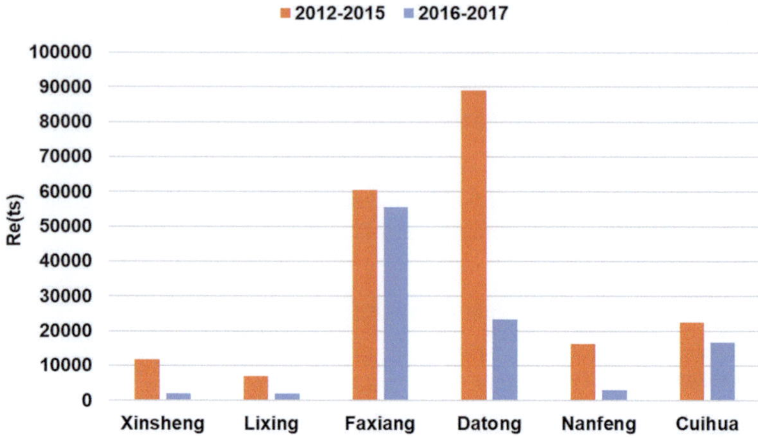

Fig. 5 The resilience values of the six villages in the study area, for Case B with the reconstruction budget model

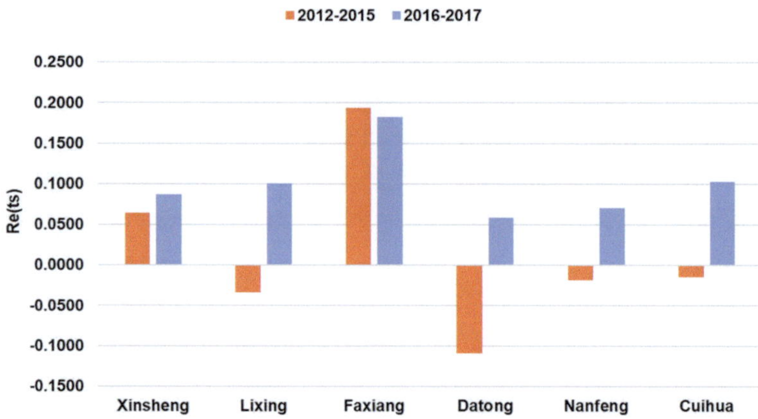

Fig. 6 The resilience values of the six villages in the study area, for Case C with the reconstruction budget model

4.2 Model Based on Population Density

For the population density model, the applied data was obtained by the best regression formula, with the largest R^2 value. The seasonal data of 2012–2017 was used as $C(t)$. In order to compare the temporal behavior of population density of each village, two periods of time, i.e., 2012–2015 and 2016–2017, were considered. The unit time of an increment Inc(t) is one season, and the total time (seasons) NI(t) is 16 seasons in 2012–2015 and 8 seasons in 2016–2017.

The results in Fig. 7 indicate Nanfeng village is obviously different from the other villages. It could be due to the high regression error of 54%, that also reflects

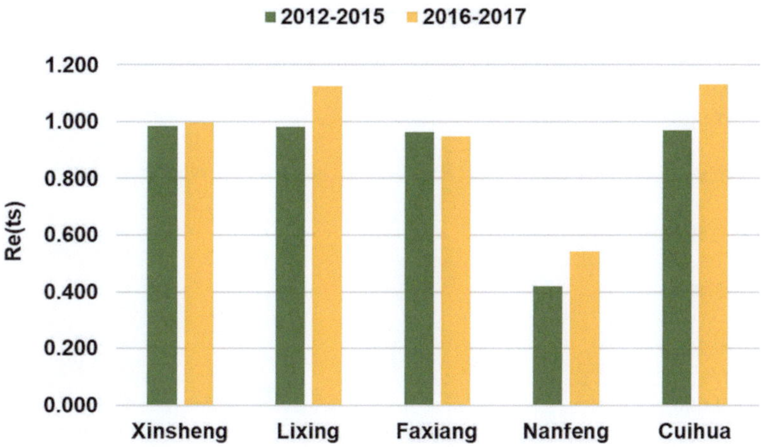

Fig. 7 The resilience values of the six villages in the study area, for Case B with the population density model

an abnormal behavior of population density. Unlike the other villages, there is no significant variation of population density in Xinsheng and Faxiang villages. Similar to electricity consumption, the population density is an indicator of economic activity. The larger of those two indicators, the larger disaster resilience.

4.3 Model Based on Electricity Consumption

Similarly, the regression formula with the highest R^2 value is also used to calculate the average electricity consumption per household for each season of 2012–2017 as $C(t)$. The unit time of an increment $\mathrm{Inc}(t)$ is one season, and the total time (seasons) $\mathrm{NI}(t)$ is 16 seasons in 2012–2015 and 8 seasons in 2016–2017. According to the results of the regression analysis, although the R^2 value is acceptable, but the average error in the performance of the forecast results is up to 68% (Faxiang Village). However, Cuihua village possesses the largest increases in population density and electricity consumption, also has the highest disaster resilience out of the six villages.

Two different interpretations are summarized according to different indicators of disaster resilience. First of all, the reconstruction budgets are as the cost of post-disaster repair, the more budget invested the more time the region needs to return to a stable state. Therefore, as a wayward evaluation value the greater the disaster resilience is lower. In addition, based on the results of Cuihua Village, the population density and electricity consumption are the highest compared with other villages, and the comparison with the calculation results of the village's reconstruction budget shows that the reconstruction budgets of the village in 2008–2018 were relatively large compared with other villages (Fig. 8).

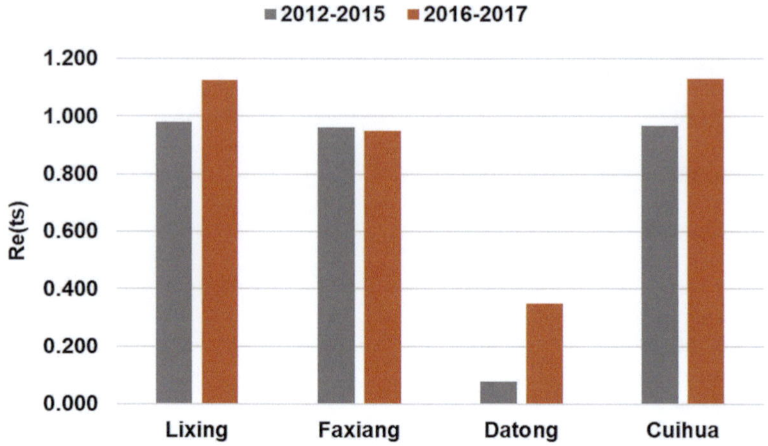

Fig. 8 The resilience values of the villages in the study area, for Case B with the electricity consumption model

5 Modify Linear Regression of Electricity Consumption

Due to the poor fit of the linear model for the average electricity consumption mentioned earlier, this study further discovered that Taiwan Power Company implemented a bi-monthly meter reading and billing system, separating residential and commercial customers since 1985. Residential households have their electricity usage recorded in odd-number months, while commercial customers have theirs recorded in even-number months. Therefore, in this study, the total electricity consumption for both residential and commercial customers was summed and averaged. Since the billing is based on meter readings taken one to two months before, for example, the actual total electricity consumption for April 2012 is calculated by adding the residential electricity consumption recorded in May 2012 and the commercial electricity consumption recorded in June 2012, and then dividing by two. By employing this improvement method, the previously observed pattern of higher consumption in odd months compared to even months was corrected. The revised average electricity consumption now exhibits a smoother trend with no significant fluctuations, aligning better with normal usage patterns.

The corrected linear regression formula, as shown in Eq. (6), has an R^2 value of 0.8, which is a significant improvement compared to the uncorrected formula (Eq. 5) with an R-squared value of 0.596.

$$\begin{aligned} Y = -0.117 * \mathrm{RI1} - 0.007 * \mathrm{RI2} - 0.015 * \mathrm{RI3} - 0.136 * \mathrm{RI4} \\ + 0.127 * \mathrm{RI5} - 0.116 * \mathrm{RI6} - 0.007 * \mathrm{RI7} + 1.034 \end{aligned} \tag{6}$$

When further comparing with the uncorrected resilience model of average electricity consumption, the maximum validation error percentage after the correction

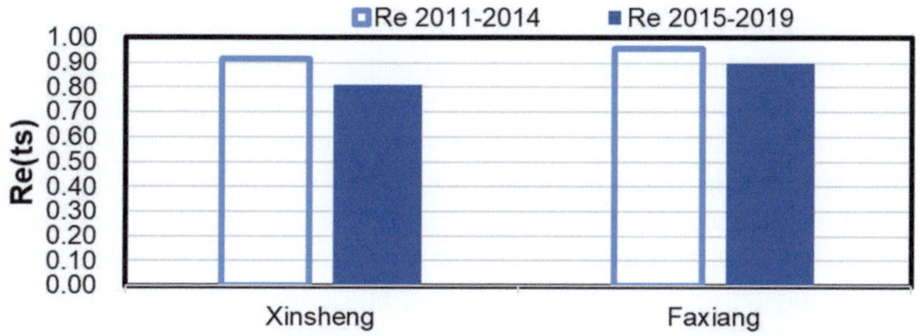

Fig. 9 The resilience values of the villages in the study area with the corrected electricity consumption model in high seasons

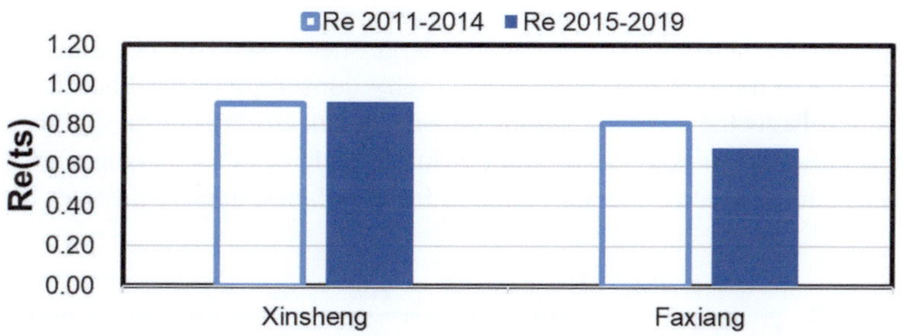

Fig. 10 The resilience values of the villages in the study area with the corrected electricity consumption model in low seasons

is reduced from 500 to 83%. From this observation, it is evident that the electricity consumption data correction method used in this section effectively improves the reliability of the landslide resilience model, resulting in a significant increase in the explanatory power of the average electricity consumption resilience model. The same improvement is observed when applying the Karen [3] formula. Figures 9 and 10 show the assessment results for average electricity consumption per household during high and low seasons.

6 Application of Landslide Resilience Model Under Extreme Rainfall Climate

Considering the impact of climate change, the extreme rainfalls predicted by the TCCIP for the mid-twentieth century and end of the century were used to calculate the susceptibility value using the previously mentioned landslide susceptibility model. Subsequently, this value will be applied to the landslide resilience model to obtain the resilience under extreme conditions.

To compare with recent resilience data, as all future extreme rainfall events occur between June and August, the second quarter of the year 2019 will be selected for comparison. By comparing the present with the future, this study assessed the impact of extreme rainfall events in the mid to late twenty-first century on resilience and plan corresponding adaptive measures to reduce disasters. Figure 11 shows the comparison of resilience between the recent and future population densities using linear regression. Figure 12 displays the comparison of resilience between the recent and future average electricity consumption per household using linear regression.

Overall, regardless of whether considering population density or average electricity consumption per household, the resilience of all villages to extreme rainfall climate is lower than the recent resilience. This implies that intense rainfall can reduce resilience and slow down the post-disaster recovery process. Further comparison of the two dependent variable resiliencies shows that the average electricity consumption per household experiences a larger decrease. This phenomenon may be attributed to the fact that average electricity consumption per household more accurately reflects the local population density. As electricity consumption tends to remain stable, a larger population would result in higher electricity consumption. On the other hand, the data collection method for population density involves dividing

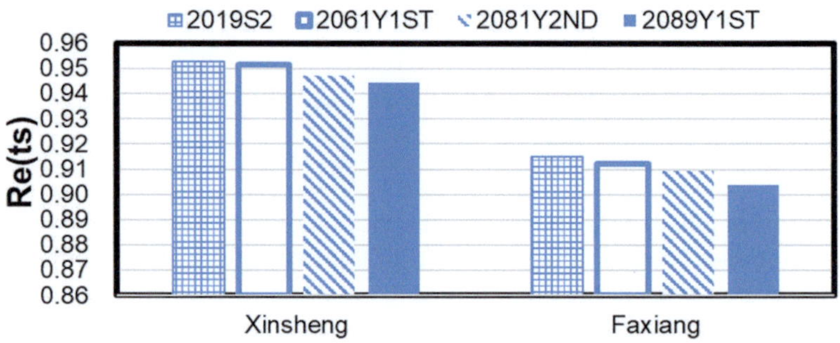

Fig. 11 Comparison of resilience of population density between recent and extreme rainfalls

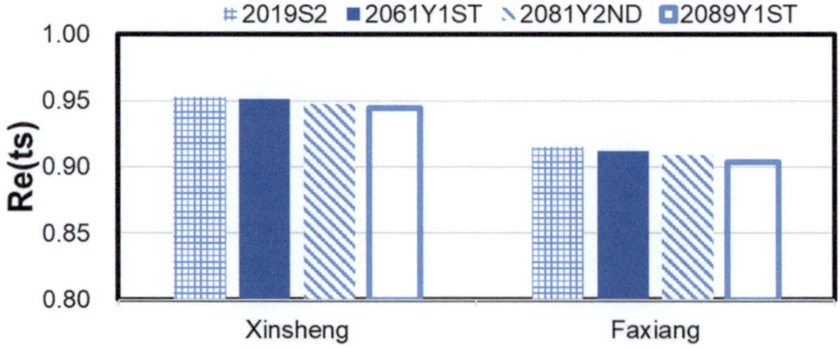

Fig. 12 Comparison of resilience of electricity consumption between recent and extreme rainfall

the registered residents in the village by the land area of the village, making it difficult to reflect temporary population during touring or harvesting seasons.

7 Conclusions and Discussions

This study adopted the upstream of Wu River in Central Taiwan as the study area, and chose the control factors by to establish the resilience model by linear regression method. In order to incorporate the concept of disaster resilience, population density, electricity consumption, and reconstruction budget were used as the indicators of resilience. And the following conclusions are drawn from the analysis results:

For the landslide susceptibility, this study adopted the Logistic Regression model of Lin [2] to produce landslide susceptibility values. Based on the analysis results using the resilience model of Karen [3], with considering reconstruction budget, population density, and electricity consumption as resilience indicators, it suggests the smaller the value calculated by the reconstruction budget represents higher resilience. On the other hand, for population density and electricity consumption, the smaller the calculated value the smaller the resilience.

After adjusting the average electricity consumption per household data, the results of the linear regression model show a better fit. When applied to extreme rainfall scenarios, the population density exhibits less variation compared to the average electricity consumption per household, indicating that average electricity consumption per household better reflects population movements and changes.

Acknowledgements This research was made possible by the financial support of the Ministry of Science and Technology (Project No. 109-2625-M-005-001) of Taiwan. We also deeply appreciate the databases and support from the Central Geological Survey project (Project No. 97-5826901000-05).

References

1. UNISDR (2015) Sendai framework for disaster risk reduction 2015–2030. In: UN world conference on disaster risk reduction, 2015 March 14–18, Sendai, Japan
2. Lin JF (2016) On the rainfall factor for the landslide susceptibility analysis in Central and Southern Taiwan areas. Thesis, National Chung-Hsing University
3. Karen SR (2011) On landslide risk, resilience and vulnerability of mountain communities in Central-Eastern Nepal. Thesis, University of Lausanne

Study on Climate Change Related Seismic Landslides of Loess Deposit

Lanmin Wang, Xingyu Ma, Kun Liu, Xiaowu Pu, Shaofeng Chai, and Ping Wang

Abstract With global warming, extreme precipitation anomalies have obviously intensified during the recent 56 years in the Loess Plateau of China, which may cause water content of loess deposit increasing sometimes. The risk of seismic landslides and large-scale mudflows in the region has been increased predominantly due to the climate change. In this paper, field investigation and exploration, in-situ rainfall tests, shaking-table tests, laboratory tests and numerical analysis were carried out to study the mechanism, characteristics, and risk assessment of rainfall related seismic landslides of loess deposit. The results shown that: (1) Extreme rainfall may saturate or wet the top layer of loess slopes within 2–3 m deep so that the shear strength of the top loess decreases dramatically. Liquefaction may trigger a mudflow in the saturated loess layer and subsidence or shallow landslides may develop in the wetting loess layer under a strong earthquake. (2) If there are vertical fissures or cracks in loess deposit of slopes, rainfall water may permeate deeper soil layer to saturate the bottom layer of a slope. The saturated bottom layer may liquefy to develop a sliding surface under the effect of strong earthquakes, along which the overburden unsaturated soil deposit would slide down for a long distance. (3) The areas with high and moderate risks of climate change related seismic landslides for the Loess Plateau under the effect of earthquakes with an exceedance probability of 10% and 2% in 50 years respectively exist in the western and southeastern plateau.

Keywords Climate change · Rainfall · Seismic landslides · Loess plateau

L. Wang (✉) · K. Liu · X. Pu · S. Chai · P. Wang
Lanzhou Institute of Seismology, CEA, Lanzhou, China
e-mail: wanglm2304@126.com

Key Lab of Loess Earthquake Engineering, CEA, Lanzhou, China

X. Ma
Institute of Engineering Mechanics, CEA, Harbin, China

H. Hazarika et al. (eds.), *Multidisciplinary Approach for Disaster Resilience and Sustainable Technologies*, Lecture Notes in Civil Engineering 642,
https://doi.org/10.1007/978-981-96-8618-6_5

1 Introduction

The Loess Plateau is located in northwestern China with an area of 440,000 km^2 and a deposit thickness ranging from tens meters to more than 500 m, where is both an earthquake-prone region and landslides-prone region. In history, more than 1.4 million people were killed by earthquakes and earthquake-triggering landslides [1]. With global warming, extreme precipitation anomalies have obviously intensified during the recent 56 years in the Loess Plateau of China (Figs. 1 and 2) [2]. Extreme precipitation anomalies have become more again since 2010s after a decreasing during 1980s–1990s. In recent years, 70% of weather stations all over the Loess Plateau observed more extreme precipitation although the rainfall in the whole plateau trends to decrease. Especially, the extreme precipitation from June to August reached 54% of the total precipitation of a summer. Such extreme continuous rainfall or torrential rain in a short time may cause water content of loess deposit increasing dramatically. The risk of seismic landslides and large-scale mudflows in the region has been increased predominantly due to the climate change. It is just such a case that the Minxian–Zhangxian 6.6 earthquake induced more than 600 landslides and a large-scale mudflow in loess deposit due to a continuous twice rainfall as much as usual in 21 days before the earthquake [3]. Most of the previous studies focused on loess landslides individually caused by rain or earthquakes [3–5], and there are few studies on loess landslides under the coupling of rainfall and earthquake. In this paper, field investigation and exploration, in-situ rainfall tests, shaking-table tests, laboratory tests and numerical analysis were carried out to study the mechanism, characteristics, and risk assessment of rainfall related seismic landslides of loess deposit. The research results have important scientific significance for preventing and mitigating such disasters in the Loess Plateau.

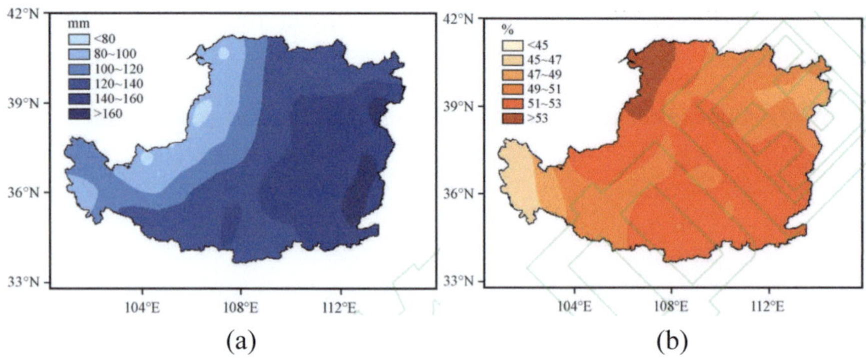

Fig. 1 Distributions of extreme precipitation (**a**) (mm) and its proportion (**b**) (%) in summer during 1961–2016

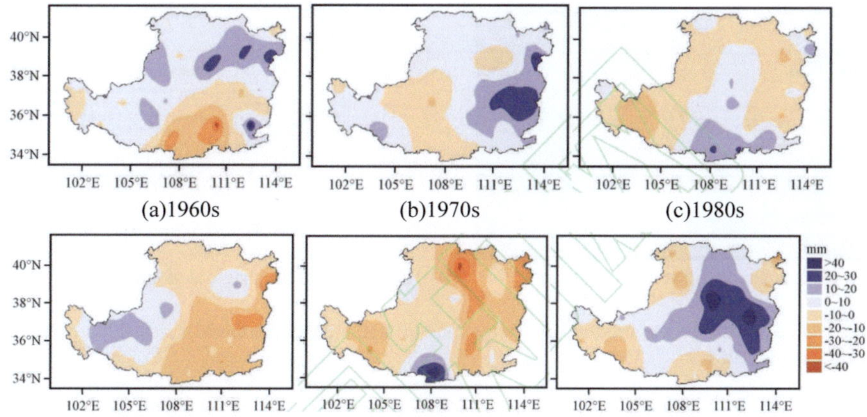

Fig. 2 Decadal extreme precipitation anomalies (mm) in summer from 1960 to 2010s

2 The Influence of Extreme Rainfall on Seismic Landslides

2.1 Field Investigation to a Mudflow at Yongguang Village

A Ms6.6 earthquake with focal depth of 20 km occurred at the junction of Minxian county and Zhangxian county, Gansu Province, China, on July 22, 2013, which caused 95 people dead, 2414 people injured and an economic loss of more than US $5 billon. The epicenter is located in the transitional zone from the Gannan Plateau to the Loess Plateau and the Longnan Mountains, where is in the northeast margin of Qinghai-Tibet Plateau. About 2330 landslides, mudflow and collapses as well as subsidence were induced in the meizoseismal zone with intensity of VIII degree and its surrounding area with intensity of VII degree. The landslide and consequent mudflow in the western Yongguang village is one of them with the biggest scale and longest runout distance (Fig. 3).

The field investigation and exploration shown that the mudflow developed in the western landslide on a loess slope with an original slope angle of 15–22° at Yongguang village, which is about 1 km long, 15–110 m wide, 6–12 m thick, 76 m high, and 420,000 m^3 in volume (Fig. 3a). The underground water table is in wind-laid loess deposit (Q_3) shown as the green line in Fig. 3b. The sliding surface developed in the interface between saturated wind-laid loess (Q_3) and the underlying moderately weathered argillaceous sandstone at the middle-up slope shown as the red line in Fig. 3b. The whole process of sliding generated by liquefaction could be divided into two stages (Fig. 4). In the first stage, the liquefaction in the deep loess layer of the slope saturated with underground water triggered a sliding of overburden unsaturated loess and saturated top loess layer in the middle-up part of the slope for about 350 m shown as the red region of Fig. 4. In the second stage, starting several minutes after

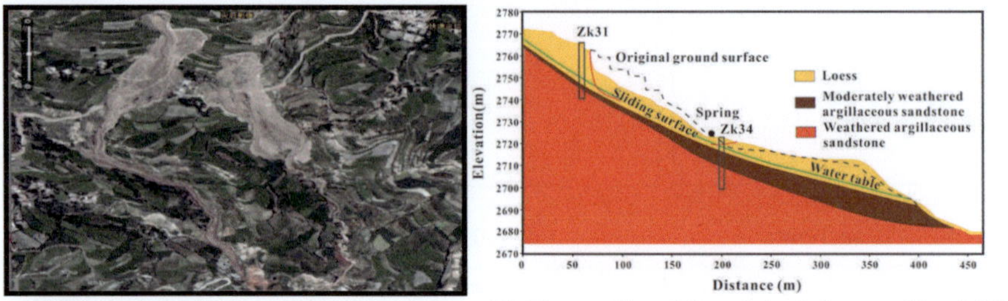

(a) loess landslides and mudflow (b) The profile of the original slope and landslide

Fig. 3 The landslides and mudflow at Yongguang village induced by the Minxian–Zhangxian 6.6 earthquake in 2013

the first stage finished, the liquefied top layer loess saturated with rainfall water flowed down along a gully of the loess mountain for about 700 m shown as the green region of Fig. 4. The mudflow is attributed to liquefaction of the top loess layer saturated with an extreme continuous rainfall in 21 days before the earthquake, which accumulated 126.3 mm twice as much as usual. Considering the amplification effect of loess mountain with a relative height of 460 m comparing with the ground motion station at Mixian County, the PGA at the landslide site is estimated as 462–715 gals based on the time histories of ground acceleration recorded at Minxian county station which is 8 km away from the landslide site (Fig. 5).

2.2 *The Artificial Rainfall Tests on the Saturated Depth of Loess Deposit*

In order to investigate the influence of extreme rainfall on the saturated depth of loess deposit, the in situ artificial rainfall tests were carried out on a loess slope of 20° in Lanzhou New District which have the same wind-laid loess deposit formed in Q_3 with that in Yongguang village. The artificial rainfall lasted five days to simulate continuous light, moderate, heavy, torrential rains in a short period which may appear in the Loess Plateau (Fig. 6a). The sensors of pore water pressure and water content were set in the slope with different depths and positions to investigate the maximum depths of saturated and wetting soil during the continuous rainfall (Fig. 6b). The observed distribution of saturated and wetting soil shown as Fig. 6c. The tested results show that the maximum depth of saturated loess layer is 2 m. If vertical fissures exist in loess deposit, the rainfall water can go down as deep as the fissures.

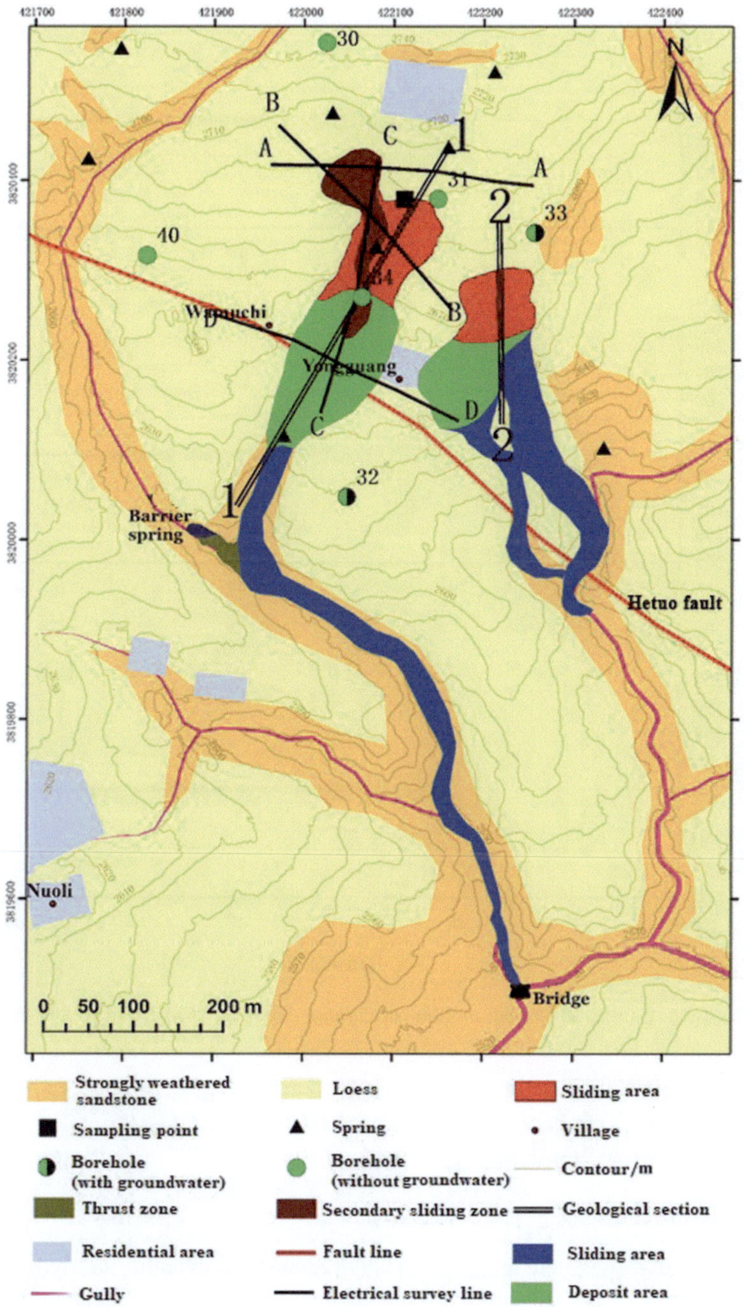

Fig. 4 The mechanism demonstration of the mudflow at the western Yongguang village

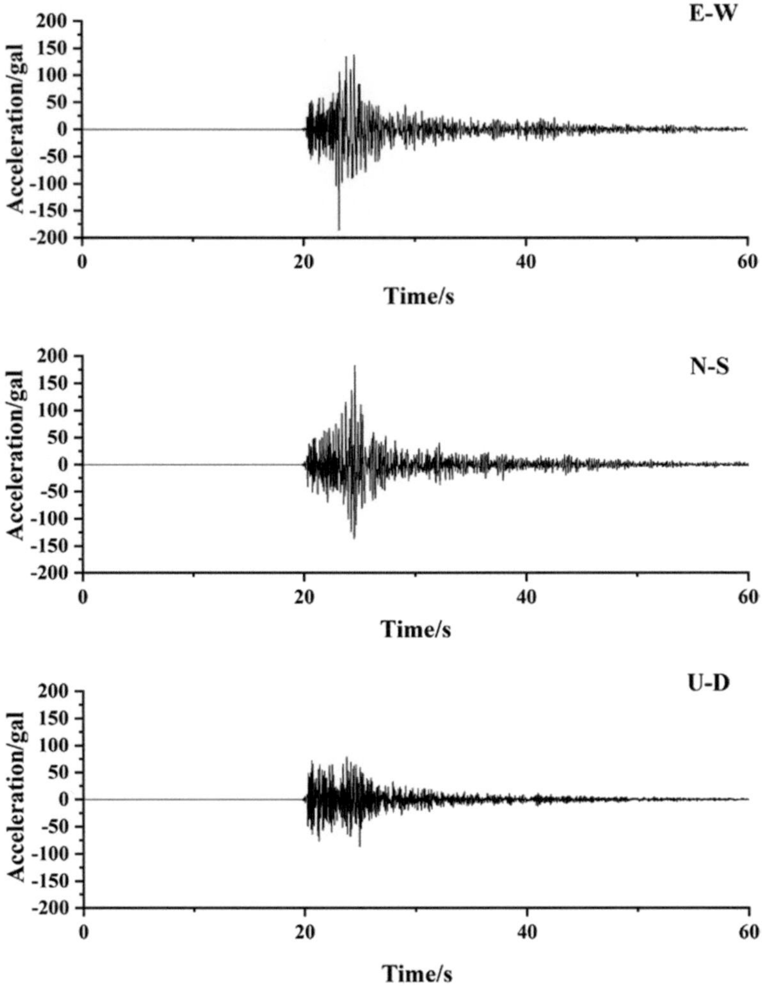

Fig. 5 The time histories of ground motion acceleration recorded at Minxian county during the earthquake

2.3 Dynamic Triaxial Tests on the Influence of Water Content on Dynamic Strength of Loess

A series of dynamic triaxial tests on undisturbed wind-laid loess specimens were performed to investigate the influence of water content on dynamic strength of loess. The loess samples were secured from the depths of 5 m and 10 m deep undisturbed loess deposit (Q_3) in Lanzhou. Table 1 presents its physical parameters, including natural density (ρ), dry density (ρ_d), natural water content (w), plastic limit (PL), plastic index (PI), and grain composition. The tests results shown that both dynamic

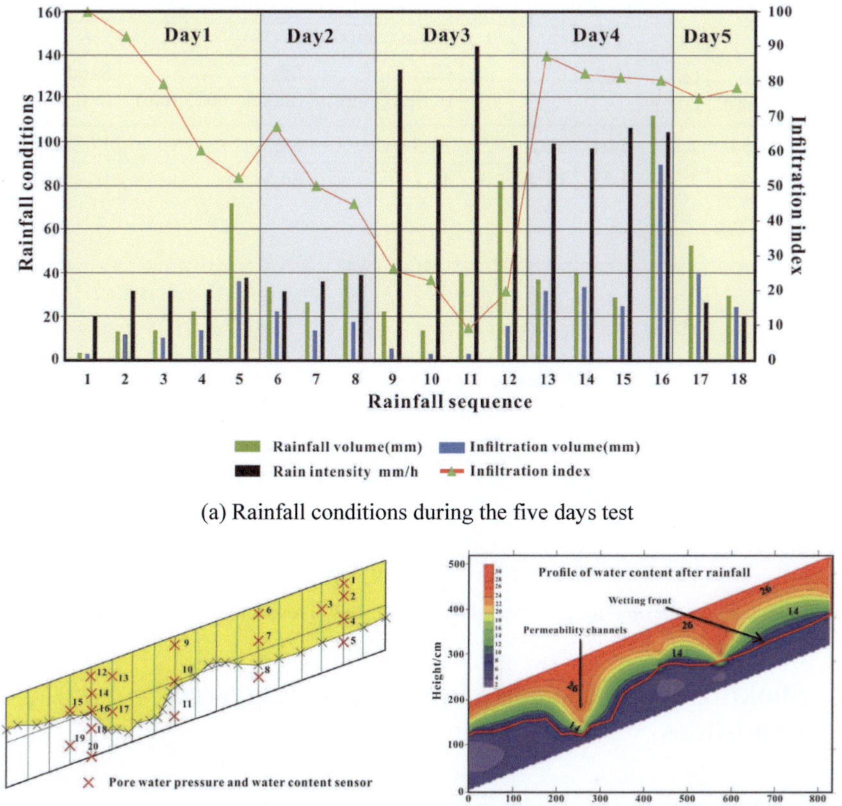

(a) Rainfall conditions during the five days test

(b)The profile of natural loess slope for the test (c) Distribution of saturated and wetting soil

Fig. 6 In-situ test of artificial extreme rainfall on a loess slope in Lanzhou New District, China

cohesion (C_d) and internal friction angle (φ_d) decrease with increasing of water content of loess specimens under equivalent sinusoidal loadings shown as Fig. 7. The empirical formulae presented in the two figures can estimate dynamic cohesion and internal friction angle of loess with different water content under the effect of earthquakes with intensities of 7 degree (10 cyclic times) and 8 degree (20 cyclic times). Thus, the influence of rainfall on seismic stability of slopes can be evaluated in terms of dynamic strength parameters of loess.

Table 1 The physical parameters of loess specimen for dynamic triaxial tests

Depth (m)	ρ (g/cm^3)	ρ_d (g/cm^3)	ω (%)	PL (%)	PI	Grain composition (%)		
						Clay (< 0.005 mm)	Silt (0.005–0.075 mm)	Sand (> 0.075 mm)
5	1.34	1.28	5.04	17.2	7.3	14.5	68.2	17.3
10	1.36	1.30	4.25	18.2	7.5	17.1	66.5	16.4

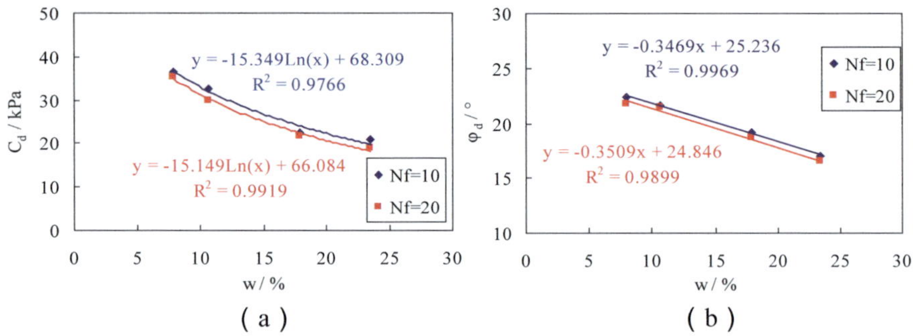

Fig. 7 Dynamic cohesion (**a**) and internal friction angle (**b**) respectively versus water content

3 Shaking Table Tests on Liquefaction-Triggered Landslides

Liquefaction-triggered landslides in loess deposit may be divided into three types of landslides in terms of position of liquefied layer, deep layer liquefaction-triggered landslides, top layer liquefaction-triggered mudflow, and joint liquefaction of both deep and top layers-triggered landslide. In the study, two series of shaking-table tests were performed for the former two types of liquefaction-triggered landslides.

3.1 Shaking Table Tests on Deep Liquefaction-Triggered Landslides

The electro-magnetic driven shaking-table tests on deep liquefaction-triggered landslide were carried out on a slope model with angle of 20° to simulate the first stage of loess landslide at the western Yongguang village (Fig. 8). The model is 300 cm long, 110 cm wide, and 100 cm high. There is 25 cm thick mudstone at the bottom and the upper part is homogenous wind-laid loess secured from the landslide site. The geometric similarity is 1:10 and physical parameters are taken as the same as the landslide site (Table 2). Before shaking, the deep loess layer with thickness of 30 cm between the mudstone and the upper unsaturated loess were saturated by

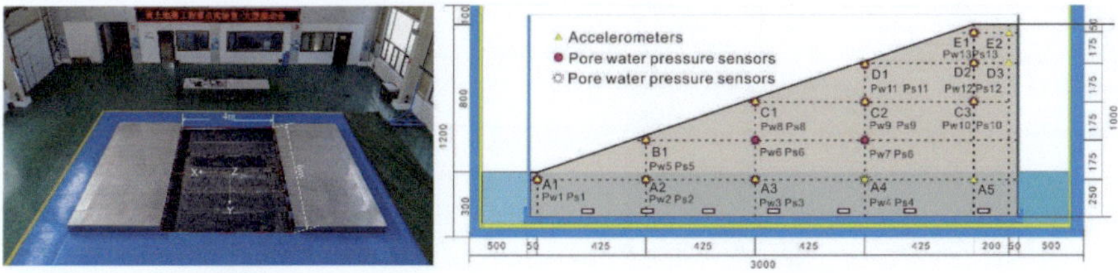

(a) (b)

Fig. 8 The 4 × 6 m electro-magnetic shaking-table (**a**) and the tested slope model with sensors (**b**)

Table 2 Loess parameters for the shaking-table test model

Physical indexes	Similarity ratio	Actual value	Target value	Test value
ω (%)	1	4.1–5.6	4.1–5.6	6
ρ/(10 kg/m^3)	1	1.31–1.34	1.31–1.34	1.32
C (kPa)	10	43.2–50.2	4.3–5	9
φ(°)	1	35–37	35–37	33

seepage of a water pipe in the mudstone. During the tests, EW-component of ground motion acceleration record at Minxian count were applied to the shaking-table with different amplitudes step by step. The test results in Fig. 9 shown that pore water pressure in the middle part of the slope increased slightly and several small fissures appeared at middle-lower part and top of the slope under the shaking PGA of 226 gals. Increasing of pore water pressure developed backward from the middle part of the slope and fissures became more and bigger under the shaking PGA of 339–452 gals. The pore water pressure increased predominantly, the sliding surface formed along the upper layer of liquefied loess, and the overburden unsaturated loess and the upper liquefied loess slip down along the sliding surface under the shaking PGA stronger than 565 gals. This mechanism is consistent with the first stage of loess landslide at the western Yongguang village induced by the Minxian–Zhangxian 6.6 earthquake in 2013.

3.2 *Shaking Table Tests on Top Liquefaction-Triggered Mudflow*

In order to simulate the mudflow at the second stage of loess landslide in the western Yongguang village, the same shaking table apparatus was used to the tests of top liquefaction-triggered mudflow. The tested slope model with sensors of acceleration, pore water pressure and soil pressure is shown in Fig. 10a. The soil parameters and similarity are the same as Table 2. Before the test, the top loess layer of 20 cm thick was saturated by artificial rainfall.

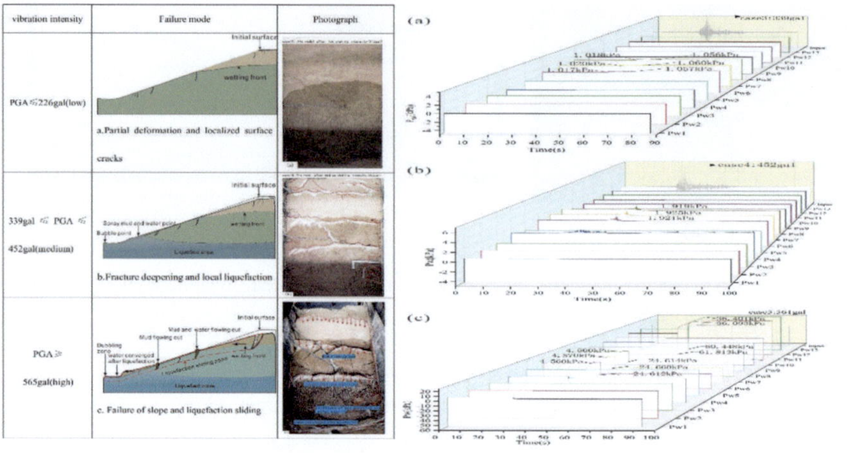

Fig. 9 Failure modes and development of pore water pressure in the loess slope

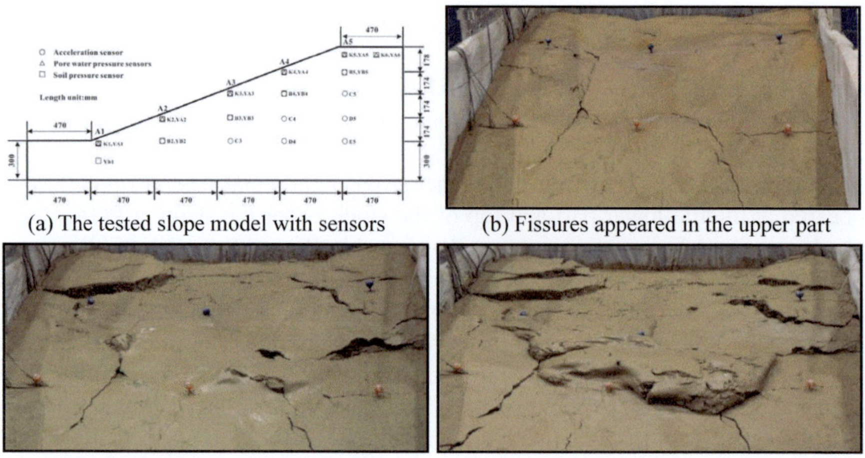

(a) The tested slope model with sensors

(b) Fissures appeared in the upper part

(b) Local liquefaction in the upper part

(d) Mudflow generating in the upper part

Fig. 10 The shaking table test model and failure process of top liquefaction-triggered mudflow

The tested results shown that liquefaction appeared at a upper middle point, several fissures developed and pore water pressure on the slope crest increased under the shaking PGA of 400 gals (Fig. 10b, Fig. 11). Liquefaction expanded the upper part of the slope, fissures enlarged into cracks and pore water pressure increased predominantly in the most part from the crest to lower part of the slope under the shaking PGA of 600 gals (Figs. 10c and 11). Mudflow was triggered at the upper part of the slope and pore water pressure increased dramatically on the whole top layer of the slope. Both the PGA triggering mudflow and the failure process strongly support the

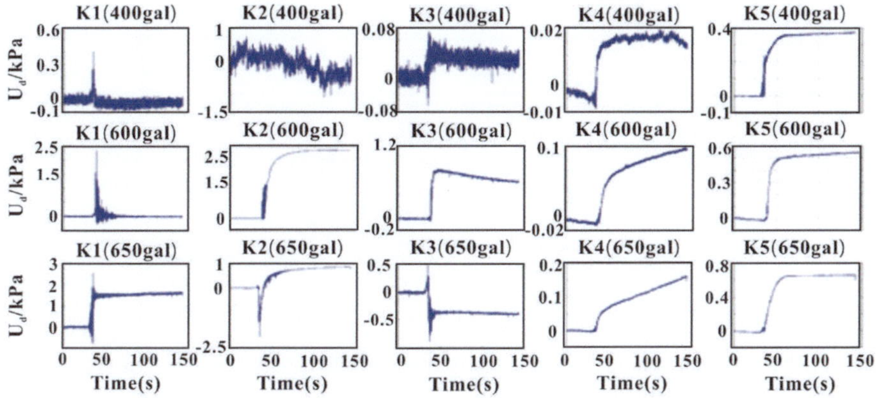

Fig. 11 Development of pore water pressure under different PGAs during the shaking table tests

mechanism of the mudflow at Yongguang village induced by Minxian–Zhangxian 6.6 earthquake mentioned in the Sect. 2.1.

4 The Risk of Climate Change Related Seismic Landslides for the Loess Plateau, China

The above-mentioned extreme precipitation was considered in the risk assessment of climate related seismic landslides for the Loess Plateau together with seismic peak ground acceleration, stratum, morphology type, slope angle and natural density. The risk is divided into four grades in terms of susceptibility of climate related seismic landslides in loess deposit shown in Table 3. The risk zonation maps of climate related seismic landslides for the Loess Plateau were respectively compiled with different probabilities of exceedance of PGA based GIS shown as Fig. 12.

Figure 12a shows that the western plateau has moderate risk, and the southern and eastern plateau have low risk under the effect of basic seismic action with an exceedance probability of 10% in 50 years. Figure 12b shows that the western plateau

Table 3 The risk grading criterion of climate related seismic landslides in loess deposit

Susceptibility	Landslide index	Characteristics of hazards
Very low	< 2.8	Seldom, stable slope
Low	2.8–3.5	Less, and no large landslides
Moderate	3.5–4.3	More, most landslides with medium scale
High	> 4.3	Many, dense and widespread, large landslides frequently and mudflow possibly

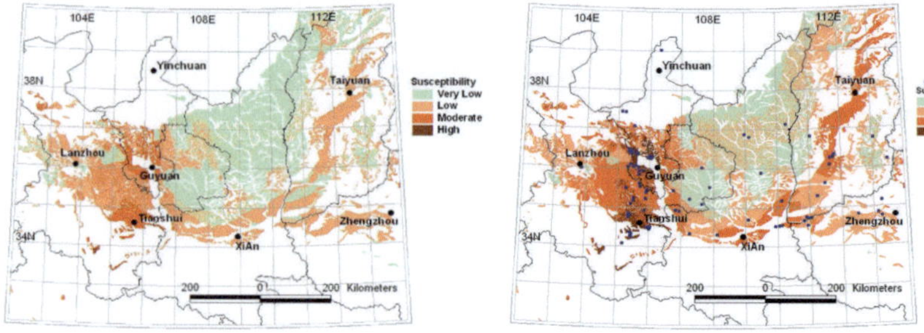

(a)10% in 50 years (b) 2% in 50 years

Fig. 12 Risk zonation maps of climate related seismic landslides in the Loess Plateau with different probabilities of exceedance of PGA

has high and moderate risk, the southern and eastern plateau have moderate risk, and the other areas have low risk under the effect of rare seismic action with an exceedance probability of 2% in 50 years. Comparing with the distribution of decadal extreme precipitation anomalies (mm) in summer from 1960 to 2010s in Fig. 2, we can see that extreme precipitation has obvious influence on seismic landslides as long as the earthquake intensity is strong enough.

5 Conclusion

(1) With global warming, extreme precipitation and continuous rainfall in several days may saturated or wetting the top layer of slopes, even deep soil if fissures exist, in the Loess Plateau, which may seriously influence seismic stability of loess slopes. In such condition, more and larger scale mudflows, landslides and inhomogeneous settlements may be triggered by a strong earthquake.
(2) Considering climate change, the areas with high and moderate risks of seismic landslides for the Loess Plateau under the effect of earthquakes with an exceedance probability of 10% in 50 years and 2% in 50 years respectively exist in the western plateau, southern and eastern plateau.

Acknowledgements The research work in this paper is financially supported by the key project of National Natural Science Foundation of China (Grant No: U1939209).

References

1. Lanmin W (2020) Mechanism and risk evaluation of sliding flow triggered by liquefaction of loess deposit during earthquakes. Chin J Geotechn Eng 42(1):1–19

2. Lu S, Hu Z, Fu C et al (2021) Characteristics and possible causes for extreme precipitation in summer over the Loess plateau. Plateau Meteorol:1–14. https://doi.org/10.7522/j.issn.1000-0534
3. Lanmin W, Zhijian W (2016) Earthquake damage characteristics of the Minxian–Zhangxian Ms6.6 earthquake and its lessons. China Earthquake Eng J 38(04):533–540
4. Wang JJ, Liang Y, Zhang et al (2014) A loess landslide induced by excavation and rainfall. Landslides 11:141–152
5. Wu LZ, Zhou Y, Sun P et al (2017) Laboratory characterization of rainfall-induced loess slope failure. CATENA 150:1–8

Extreme Rainfall Caused by Climate Change and Its Effects of on Debris Flow Hazard in Taiwan

Meei-Ling Lin and Te-Wei Chen

Abstract Due to the trend of climate change, the extreme rainfall event could last for a long duration with a high intensity. Such extreme rainfall event is likely to trigger severe landslides and debris flows. Typhoon Kalmaegi with a high rainfall intensity and an accumulated rainfall of more than 1000 mm caused severe landslides and debris flow hazard in southern Taiwan in 2008. Typhoon Morakot with different rainfall pattern also caused extensive landslides and debris flow hazard in southern Taiwan in 2009 with an accumulated rainfall of more than 3000 mm. The study area includes five debris flow torrent basins in southern Taiwan with debris flow triggered in both events. Field investigation and numerical simulations of the debris flow were conducted on the five basins of the two typhoon events. The analysis results show that the higher accumulated rainfall leads to larger debris deposition volume, inundated area, and deposition depth, which consists with the field investigation results, and suggests a higher impact to debris flow basin. The effect of high accumulated rainfall is more significant for small basin, and no correlation was found for rainfall intensity. A linear relation can be found for deposition volume versus basin area caused by extreme events, and the data were higher compared to worldwide dataset by Marchi et al. (2018). It is suggested that the extreme events might not fit the general trend, and properly estimating debris volume related hazard would require more comprehensive consideration of the precipitation characteristics. The high accumulated rainfall induced by climate changes would cause more severe hazard and should be taken into account for mitigation strategy.

Keywords Debris flow · Extreme rainfall · Debris deposition volume · Affected area

M.-L. Lin (✉)
National Taiwan University, Taipei, Taiwan
e-mail: linml@ntu.edu.tw

T.-W. Chen
Sinotech Engineering Consultants, Taipei, Taiwan

H. Hazarika et al. (eds.), *Multidisciplinary Approach for Disaster Resilience and Sustainable Technologies*, Lecture Notes in Civil Engineering 642,
https://doi.org/10.1007/978-981-96-8618-6_6

1 Introduction

With steep terrains and fragile geological conditions in Taiwan, the heavy precipitation often causes landslides, which could trigger debris flow subsequently. Due to the trend of climate change, both the accumulation and intensity of rainfall tend to become quite severe and different from previous precipitation records. Dore [3] synthesized from literatures and observed that wet areas become wetter, and dry and arid areas become more so. Typhoon Kalmaegi in 2008 and Typhoon Morakot in 2009 both brought high intensity and long duration rainfall but with different rainfall patterns. The two typhoons resulted in huge accumulated rainfall in southern Taiwan and both induced severe landslide and debris flow hazard. Five debris flow torrent basins with recurring debris flows in both typhoon events are selected to study the effects of different rainfall patterns on debris deposition conditions, which are important factors of debris flow impact estimation and hazard mitigation strategy. Field investigations were conducted to identify the affected area, characteristics of debris flow torrents, and debris deposition volumes were estimated where available. Numerical simulations of debris flow in the five basins were conducted for the two events, and debris deposition conditions were determined. The simulated deposition volumes and inundated areas were compared to the field data, and the relationship between the total accumulated rainfall and deposition volume were developed.

2 Study Area and Materials

2.1 Study Area

The study area is in Tainan and Kaohsiung Cities in southern Taiwan that being struck by the heavy rainfall of both typhoons. The study area includes five debris flow basins that debris flows were triggered in both Typhoon Kalmaegi and Typhoon Morakot events. The five debris flow torrents were identified as Tainan Debris Flow Torrent No. 048 (TNDF048), Kaohsiung Debris Flow Torrent No. 054 (KSDF054), Kaohsiung Debris Flow Torrent No. 064 (KSDF064), Kaohsiung Debris Flow Torrent No. 072 (KSDF072), and Kaohsiung Debris Flow Torrent No. 073 (KSDF073), respectively. Figure 1 shows the study area and locations of the five debris flow torrent basins, and Fig. 2 illustrates the geology map of the study area.

TNDF048 is in the Nanhua District, Tainan City, with the highest elevation in the basin of about 650 m, and the lowest elevation of about 200 m a.s.l., and the basin area is 325,782m^2. The main geological formation of TNDF048 basin is Peileo shale (Pa) which is at the age of Pliocene to Pleistocene and is shale and sandy shale. Other geological formations include Changchihkeng Formation (Cc) which is at the age of Miocene with sandstone, sandstone interbedded with shale, and Tangenshan Sandstone (Tn), which is at the age of Miocene to Pliocene with thick-bedded massive sandstone and muddy sandstone.

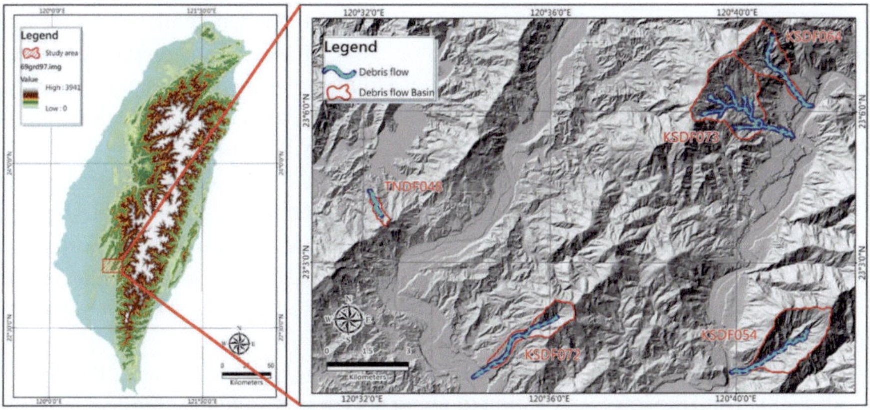

Fig. 1 Study area in Kaohsiung and Tainan cities

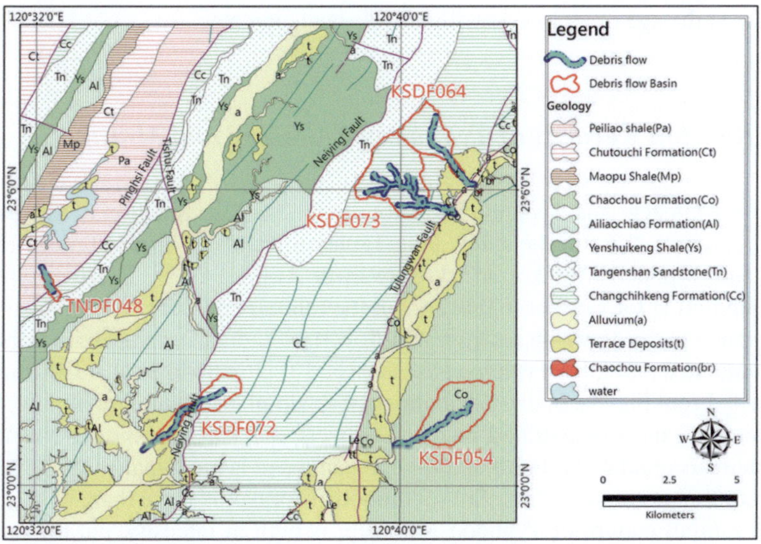

Fig. 2 Geology map of the study area

KSDF054 is in the Liugui District, Kaohsiung City, with the highest elevation of about 1340 m, and the lowest elevation of about 300 m a.s.l., and the area of basin is 4,419,285 m^2. The main geological formation of KSDF054 basin is Peiliao shale (Pa), which is at the age of Middle Miocene and is argillite and/or slate intercalated with sandstone lenses.

KSDF064 situates across the Liugui District and Taoyuan District, Kaohsiung City, with the highest elevation of about 1246 m, and the lowest elevation of about 370 m a.s.l., and the area of basin is 2,751,107 m^2. The main geological formation is

Changchihkeng Formation (Cc), which is at the age of Miocene with sandstone and sandstone interbedded with shale. The geological formation in the downstream area is Terrace Deposits (t), which is at the age of Holocene with gravel, sand, and clay.

KSDF072 is in the Shanlin District, Kaohsiung City, with the highest elevation of about 950 m and the lowest elevation of about 200 m a.s.l., and the area is 2,108,738 m^2. The main geological formation in the upstream area is Changchihkeng Formation (Cc), which is at the age of Miocene and is sandstone and sandstone interbedded with shale. Geological formation in the midstream area is Ailiaochiao Formation (Al), which is at the age of Pliocene with thin alternation of siltstone and shale. Geological formation in the downstream area is Terrace Deposits (t), which is at the age of Holocene with gravel, sand, and clay.

KSDF073 is in the Liugui District, Kaohsiung City, with the highest elevation of about 1300 m and the lowest elevation of about 450 m a.s.l., and the area of KSDF073 basin is 5,307,123 m^2. The main geological formation is Changchihkeng Formation (Cc), which is at the age of Miocene with sandstone and sandstone interbedded with shale. The geological formation at the estuary of KSDF073 is Terrace Deposits (t), which is at the age of Holocene with gravel, sand, and clay.

2.2 *Rainfall Records of Typhoons Kalmeagi and Morakot*

Typhoon Kalmaegi and Typhoon Morakot are the two precipitation events causing recurring debris flows in the study area. Typhoon Kalmaegi struck Taiwan in July, 2008, and Typhoon Morakot struck Taiwan in August, 2009. The two typhoons made landfall at the northeast Taiwan, but the rainfall concentrated in the southern Taiwan, due to the drawn-in southwest airflow from Indian Ocean. The distributions of accumulated rainfall of Typhoon Kalmaegi and Typhoon Morakot are illustrated in Fig. 3a, b. From Fig. 3a, b, the distributions of accumulated rainfall of both typhoons were similar, and concentrated in the southwest and part of central Taiwan. The maximum accumulated rainfall of Typhoon Kalmaegi is more than 1000 mm, while the maximum accumulated rainfall of Typhoon Morakot is more than 3000 mm.

The rainfall pattern of Typhoon Kalmaegi is high in intensity, but the duration is not very long as shown by the rainfall records of the Jiaxian rain-gauge station in Fig. 4a. The maximum rainfall intensity of Typhoon Kalmaegi was 150 mm/hr, and the total accumulative rainfall reached 892 mm with a duration of about one-half day. Figure 4b shows the rainfall record of Typhoon Morakot at Jiaxian rain-gauge station. The maximum rainfall intensity of Typhoon Morakot was 94.5 mm/hr, and the total accumulative rainfall was up to 1,867.5 mm with a long duration of about three days. The maximum rainfall intensity of Typhoon Kalmaegi was larger than that of Typhoon Morakot, while the accumulated rainfall of Typhoon Morakot was much larger than that of Typhoon Kalmaegi due to the long rainfall duration. Trenberth [14] found that the water-holding capacity of air increases with increasing temperature, which might have significant influence on precipitation and lead to more intense precipitation [14]. The climate change tendency suggests the extreme

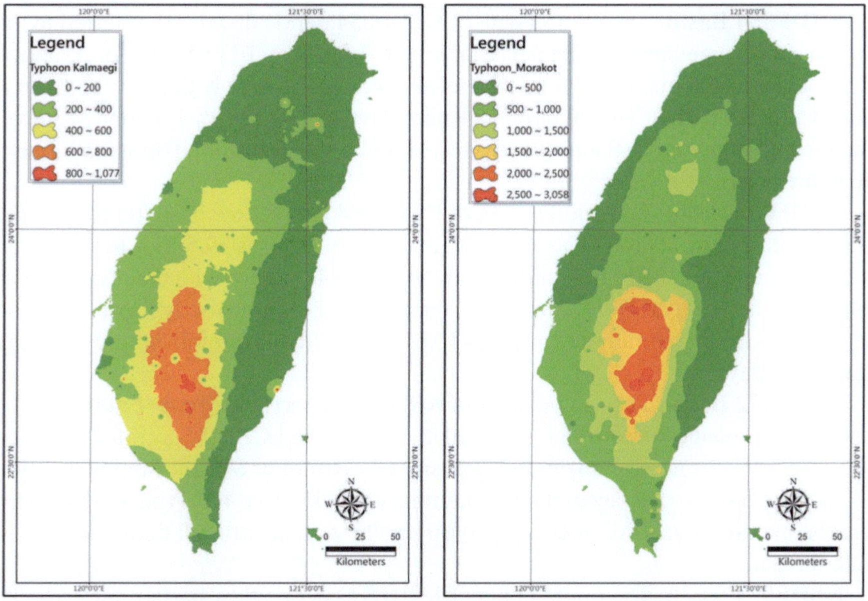

(a) (b)

Fig. 3 Distributions of the accumulated rainfall of **a** Typhoon Kalmaegi, and **b** Typhoon Morakot

weather with increase in rainfall intensity and accumulated rainfall, however, the rainfall characteristics are very different for the two events.

TNDF048 Basin. The rainfall data of TNDF048 debris flow basin referred to the Beiliau rain-gauge station. The accumulative rainfall was 2045 mm for typhoon Kalmaegi, and the maximum rainfall intensity was 143.5 mm/hr with the rainfall duration of about two days. The accumulative rainfall was 2472 mm, and the maximum rainfall intensity was 91.5 mm/hr with a duration of more than three days of Typhoon Morakot.

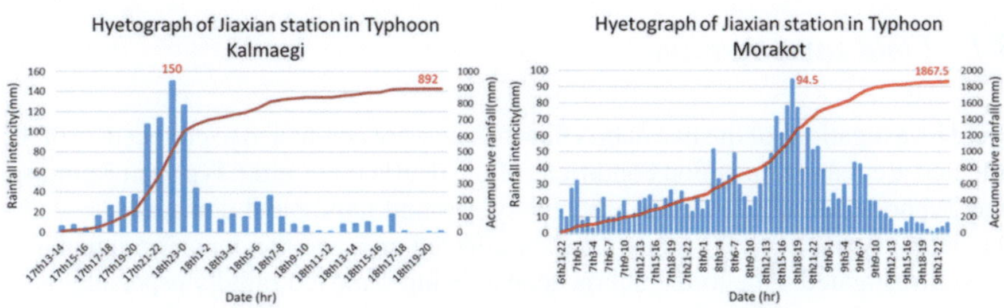

(a) **(b)**

Fig. 4 Precipitation records at Jiaxian rain-gauge station of **a** Typhoon Kalmaegi event, and **b** Typhoon Morakot event

KSDF054 Basin. The rainfall data of KSDF054 debris flow basin referred to the Shinfa rain-gauge station. The accumulative rainfall was 1039.5 mm for Typhoon Kalmaegi, and the maximum rainfall intensity was 148 mm/hr with the rainfall duration of about two days. The accumulative rainfall was 2417 mm for Typhoon Morakot, and the maximum rainfall intensity was 103 mm/hr with the rainfall duration of more than three days.

KSDF064 Basin. The rainfall data of KSDF064 debris flow basin referred to the Shinfa rain-gauge station. The accumulative rainfall was 1039.5 mm for Typhoon Kalmaegi, and the maximum rainfall intensity was 148 mm/hr with the rainfall duration of about two days. The accumulative rainfall was 2417 mm for Typhoon Morakot, and the maximum rainfall intensity was 103 mm/hr with the rainfall duration of more than three days.

KSDF072 Basin. The rainfall data of KSDF072 debris flow basin referred to the Yuemei rain-gauge station. The accumulative rainfall was 1282.5 mm for Typhoon Kalmaegi, and the maximum rainfall intensity was 76 mm/hr with the rainfall duration of about two days. The accumulative rainfall was 2424 mm for Typhoon Morakot, and the maximum rainfall intensity was 80 mm/hr with the rainfall duration of more than three days.

KSDF073 Basin. The rainfall data of KSDF073 debris flow basin referred to the Shinfa rain-gauge station. The accumulative rainfall was 1039.5 mm for Typhoon Kalmaegi, and the maximum rainfall intensity was 148 mm/hr with the rainfall duration of about two days. The accumulative rainfall was 2417 mm for Typhoon Morakot, and the maximum rainfall intensity was 103 mm/hr with the rainfall duration of more than three days.

The accumulated rainfall of Typhoon Morakot appeared to be higher than that of Typhoon Kalmaegi with a longer duration, while the maximum rainfall intensity of Typhoon Kalmaegi was higher than that of Typhoon Morakot in almost all five debris flow basins.

3 Research Methods

3.1 Field Investigation

Field investigations of the five debris flow basins were conducted to collect data of the affected areas and deposition conditions of the debris flows after Typhoon Kalmaegi and Typhoon Morakot events, respectively. Ground failure and damages caused by the debris flows, as well as the extend of affected area and the deposition thickness were investigated. Due to the debris deposit being removed rapidly especially after Typhoon Kalmaegi, the affected area and deposition thickness were identified based on field mud traces and damages caused by the debris flow, and interview of residents when available. Field survey was conducted when feasible to estimate debris volume after Typhoon Morakot.

3.2 Numerical Simulation of Debris Flow

Hürlimann et al. [4] used the one-dimensional and two-dimensional numerical models to simulate the debris flows, and used the thickness of the deposition and the flow velocity to classify the potential of the debris flow [4]. Rickenmann [11] stated that the potential initiation zones and sediment sources are the key elements for conducting model simulation with FLO-2D [11], which provided the overflow point of the debris flow. We adopted the commercially available program FLO-2D to simulate the debris flow condition, which provided the deposition volumes and inundated areas of the five debris flow basins. FLO-2D is a two-dimensional rheological model [9] approved by the Federal Emergency Management Agency (FEMA) with accountable results.

To conduct the simulation, we collected data including disaster history, topography, geology data, and rainfall hydrographs of the five basins. The major parameters of the FLO-2D analysis are Bingham yield stress and Bingham fluid viscosity, which the simulation results are very sensitive to. Lin et al. [6] and Sosio et al. [13] studied the parameter sensitivity to optimize parameters for the FLO-2D analysis [6, 13]. The computation procedures start with inputting rainfall hydrograph to derive the flood hydrograph using the hydrologic model in the FLO-2D program, and the debris flow hydrograph could be determined. Flow simulation is then conducted adopting flooding model in the FLO-2D, and the results of deposition area, thickness of deposition, and flow velocity were obtained. The simulation procedures referred to Lin et al. [7], and the hydrographs of the rainfall records of the five basins were adopted for Typhoons Kalmaegi and Morakot. The rheological and volume concentration parameters were assigned and the flooding model in the FLO 2D was then conducted to obtain the final deposition area and volume. This research referred to Chao [2] and Peng [10] for the initial input rheological and volume concentration parameters as listed in Table 1, where Bingham yield stress and Bingham Viscosity referred to Sosio et al. [13] and Hürlimann et al. [4], Manning's coefficient referred to Lin et al. [6] and O'Brien [9], and the volume concentration was set to be 0.65 for gravelly material. The parameters were then modified so that the simulated inundated area became comparable to field affected area data for each debris flow event.

Table 1 Initial parameters of FLO-2D analysis

Parameter		Value
Bingham yield stress, τ_y	α_1 β_1	0.01 13.936
Bingham viscosity, η	α_2 β_2	0.01 21.35
Manning's coefficient, n		0.2
Specific gravity, G_s		2.65
Resistance parameters for Laminar flow, k		2285
Volume concentration, C_v		0.65

4 Results of Deposition Volume and Affected Area

4.1 Field Investigation

TNDF048 Basin. Shallow landslides and erosions occurred in the TNDF048 debris flow basin leading to debris flow that blocked the drainage culvert and overflowed into and deposited around local community during Typhoon Kalmaegi. Figure 5a shows the field investigation results in Typhoon Kalmaegi, and some deposition could be observed in the figure with part of deposition removed to clear up the torrent channel. In Typhoon Morakot, the landslides occurred in Typhoon Kalmaegi were reactivated and enlarged and new landslides also occurred. Width of the TNDF048 debris flow torrent channel extended to about 20 m. The landslide mass blocked torrent channel and culvert, and debris flow flooded the surrounding area. Previous remediation facilities were buried, and the deposition volume and the affected area became much larger than those during Typhoon Kalmaegi. Figure 5b shows the changes of TNDF048 debris flow torrent, and debris deposited around and buried the community houses. The thickness of the debris deposition was about one story high (3 m high). The estimated deposition volume in the affected area was about 29,000 m^3, and the landslide area in the TNDF048 basin was about 17,000 m^2.

Fig. 5 Field conditions of TNDF048 debris flow basin during investigation: **a** after Typhoon Kalmaegi, and **b** after Typhoon Morakot

KSDF054 Basin. Landslides occurred and triggered debris flow in the KSDF054 debris flow basin during Typhoon Kalmaegi. Debris flow filled the drainage path beneath the Jioutan Bridge across the KSDF054 debris flow torrent, and debris blocked and covered the Jioutan bridge. Figure 6a shows the field condition after Typhoon Kalmaegi, and it could be found that the bridge was damaged and the channel was about 50 m wide. In Typhoon Morakot, due to the huge accumulative rainfall, more landslides occurred and debris deposition blocked the drainage path and damaged bridge foundation. Figure 6b showed that the KSDF054 debris flow torrent channel was wider than in Typhoon Kalmaegi, and the foundation of the Jioutan Bridge was damaged. The revetment of the torrent channel was scoured by the debris flow. The estimated deposition volume of KSDF054 debris flow was about 310,000 m^3, and the landslide area in the KSDF054 basin was about 100,000 m^2.

KSDF064 Basin. Many landslides occurred in KSDF064 debris flow basin causing debris flow which damaged Chianguo Bridge mid-stream of torrent in Typhoon Kalmaegi. The width of the torrent channel was 10 m before Typhoon Kalmaegi, but it extended to about 50 m after typhoon. Figure 7a shows field conditions during investigation after Typhoon Kalmaegi, and it could be observed that the

(a)

(b)

Fig. 6 Field conditions of KSDF054 debris flow basin during investigation: **a** after Typhoon Kalmaegi, and **b** after Typhoon Morakot, the debris deposit was much thicker than in Typhoon Kalmaegi, and the bridge foundation was scoured and exposed

Fig. 7 Field conditions of KSDF064 debris flow basin during investigation: **a** after Typhoon Kalmaegi, and **b** after Typhoon Morakot

confluence of KSDF064 torrent was filled by debris, which damaged the connecting road and the Chianshan No. 2 Bridge. More landslides occurred during Typhoon Morakot compared to that during Typhoon Kalmaegi, and the width of the torrent channel became about 100 m wide. Debris flow damaged the Chianshan No. 2 Bridge and huge volume of debris deposited in the torrent. From Fig. 7b, it could be found that the confluence of the KSDF064 torrent became much wider and with much more debris deposition volume than in Typhoon Kalmaegi in the estuary. The estimated deposition volume in KSDF064 debris flow affected area was about 805,800 m^3, and the landslide area in the KSDF064 basin was about 586,200 m^2.

KSDF072 Basin. Many new landslides occurred and causing debris flow in the KSDF072 debris flow basin during Typhoon Kalmaegi. Figure 8a shows the field conditions after Typhoon Kalmaegi, and it could be observed that the midstream KSDF072 torrent was filled by debris deposition. Many new landslides occurred during Typhoon Morakot due to the large accumulated rainfall comparing to that of Typhoon Kalmaegi. The landslide volumes transported into the torrent and scoured the channel banks for a length of about 60 m. Debris flow damaged the Huoshan Bridge and the huge deposition volume flooded the community houses and deposited at the confluence. From Fig. 8b, it could be found that the midstream channel of the KSDF072 debris flow torrent became wider than that in Typhoon Kalmaegi. Figure 8c shows the debris hit the community houses in the affected area, and the deposition height was higher than one story (3 m). The estimated deposition volume of KSDF072 debris flow in the affected area was about 766,500 m^3, and the landslide area in the KSDF072 basin was about 123,200 m^2.

KSDF073 Basin. Many new landslides and torrent channel scour occurred upstream of KSDF073 debris flow basin leading to debris flow during Typhoon

Fig. 8 Field conditions during investigation of KSDF072 debris flow basin: **a** midstream KSDF072 torrent after Typhoon Kalmaegi, **b** midstream KSDF072 river after Typhoon Morakot, and **c** debris deposition in the affected area after Typhoon Morakot

Kalmaegi. Figure 9a shows the field conditions during investigation after Typhoon Kalmaegi. It could be observed that the stream channel was filled by debris deposition near the Chianshan No. 1 Bridge at the confluence of torrent and mainstream. The landslides occurred in Typhoon Kalmaegi recurred and enlarged during Typhoon Morakot. Large landslide mass transported into the torrent channel and scoured the torrent bank. Debris flow flood over the Chianshan No. 1 Bridge and blocked the connecting road, and huge amount of debris deposited in the KSDF073 torrent channel. The debris deposition covered the whole Chianshan No. 1 Bridge and the connecting road with the deposition volume much larger than that in Typhoon Kalmaegi as shown in Fig. 9b. The estimated deposition volume in the affected area was about 530,000 m^3, and the landslide area in the KSDF073 basin was about 200,000 m^2.

4.2 Simulation Result

The numerical simulations of debris flow in the five debris flow basins were conducted for Typhoon Kalmaegi and Typhoon Morakot events. The simulations were based

Fig. 9 Field conditions during investigation of KSDF073: **a** after Typhoon Kalmaegi, and **b** after Typhoon Morakot

on the rainfall records adopted for each basin during the two typhoon events, and parameters were adjusted referring to affected area from field investigation for validation. The simulated results were presented using deposition depth contour on the same raster to facilitate comparisons of Typhoon Kalmaegi and Typhoon Morakot events.

TNDF048 Basin. Figure 10 shows the simulated results of inundated area and deposition depth of TNDF048 basin. The over-flow position of TNDF048 torrent was indicated as starting point of green area in the southern part of Fig. 10a for Typhoon Kalmaegi. The deposition depth is indicated in color in the figure and debris deposited along the TNDF048 channel and buried the community houses. The maximum deposition depth from simulation was about 2.48 m and lower than 2 m in most other area. Comparing to the field investigation, damages caused by the debris flow developed along and blocked the TNDF048 torrent channel, which consisted with the simulation results. Figure 10b shows the simulation result of Typhoon Morakot, and the deposition depth was higher than Typhoon Kalmaegi simulation. The maximum deposition depth was about 10.58 m, and more debris deposited near the over-flow point. The inundated area distributed along the TNDF048 torrent, but the flooded area extended to both sides of the channel and west part of community. The results were quite consistent with the damages and debris distribution caused by Typhoon Morakot in field, and appeared to cover a much larger inundated area with thicker deposit than that by Typhoon Kalmaegi.

KSDF054 Basin. The inundated area and deposition depth by KSDF054 debris flow torrent from simulation were shown in Fig. 11. Figure 11a shows the inundated area caused by Typhoon Kalmaegi, and it was observed that the debris deposited along

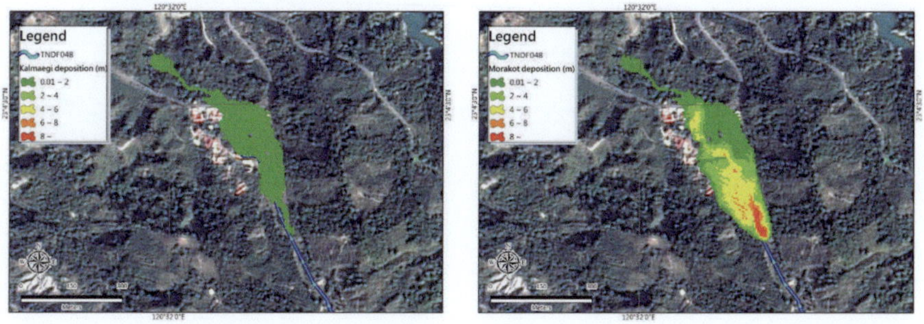

(a) (b)

Fig. 10 Simulated debris deposition of TNDF048 torrent caused by: **a** Typhoon Kalmaegi, and **b** Typhoon Morakot (base aerial photograph was taken after Typhoon Morakot)

the KSDF054 torrent channel, and concentrated at the confluence near the Jioutan Bridge. Debris deposition blocked the KSDF054 torrent channel and damaged the foundation of the bridge in field, and its distribution conformed well with the numerical simulation results. The maximum simulated deposition depth was 26.79 m in the main stream of Launong River, and debris flooded the surrounding area from confluence to the Launong River. Figure 11b shows the simulation results of Typhoon Morakot, and the inundated area was slightly larger than that in Typhoon Kalmaegi, but the deposition depth and volume were much larger than those in Typhoon Kalmaegi. The maximum simulated deposition depth was 32.35 m with high deposition depth near the confluence of KSDF054 debris flow torrent to Launong River. The deposition area in KSDF054 torrent was confined by the valley terrain, and large volume of debris flowed into the mainstream Launong River. The inundated areas caused by Typhoon Kalmaegi and Typhoon Morakot were about the same and both reached opposite riverbank of Launong River. Comparing to the field investigation, the foundation of the Jioutan Bridge was damaged in Typhoon Morakot, and the deposition depth near the bridge was larger than that in Typhoon Kalmaegi, and the numerical simulations consisted well with the field investigation results.

KSDF064 Basin. The inundated area and deposition depth by KSDF064 debris flow torrent from simulation were shown in Fig. 12. Figure 12a shows the simulation results of Typhoon Kalmaegi, and it was found that the debris deposited from the overflow point of KSDF054 torrent and concentrated along KSDF064 torrent channel. The maximum deposition depth was near the mainstream Launong River. Field investigation showed that the debris filled the KSDF064 torrent channel near confluence, which conformed with the numerical simulation well. The maximum simulated deposition depth was about 15.79 m at the mainstream Launong River, and debris flooded around the KSDF064 torrent channel to the Launong River. Figure 12b shows the simulation results of Typhoon Morakot, in which the inundated area and deposition depth were larger than those in Typhoon Kalmaegi. The maximum simulated deposition depth was about 21.87 m, and large depth occurred at the confluence of KSDF064 torrent. The deposition at the confluence damaged the

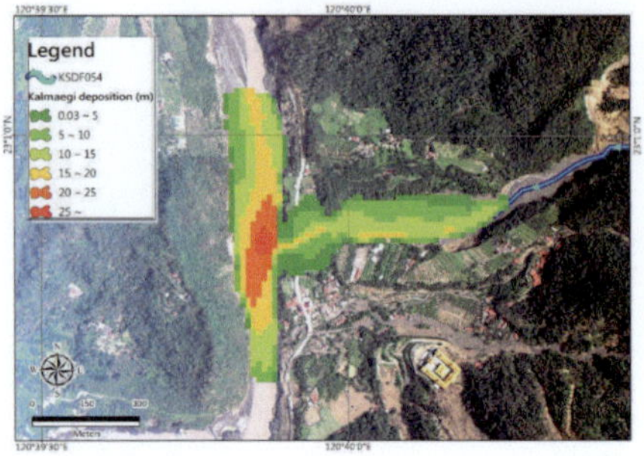

(a) (b)

Fig. 11 Simulated debris deposition of KSDF054 torrent caused by: **a** Typhoon Kalmaegi, and **b** Typhoon Morakot (base aerial photograph was taken after Typhoon Morakot)

Chianshan No. 2 Bridge, and debris flowed into mainstream of Launong River and reached the opposite riverbank. The inundated area of KSDF064 torrent in Typhoon Morakot was larger than that in Typhoon Kalmaegi, and the simulation results were quite consistent with the field investigation results.

KSDF072 Basin. The inundated area and deposition depth of KSDF072 debris torrent from simulation were as shown in Fig. 13. Figure 13a shows the simulation results of Typhoon Kalmaegi, and debris deposited from the overflow point to the confluence of KSDF072 torrent. The deposition concentrated along the KSDF072 torrent channel with the maximum deposition depth near the confluence, and some debris flowed into the mainstream Chishan River. Compared with the field investigation, the distribution of debris deposition conformed with the numerical simulation well. The maximum simulated deposition depth was about 13.15 m near the overflow point. Figure 13b shows the simulation results of Typhoon Morakot, and it was observed that the inundated area with deposition depth large than 4 m was much larger than that in Typhoon Kalmaegi, and the deposition depth was in general thicker

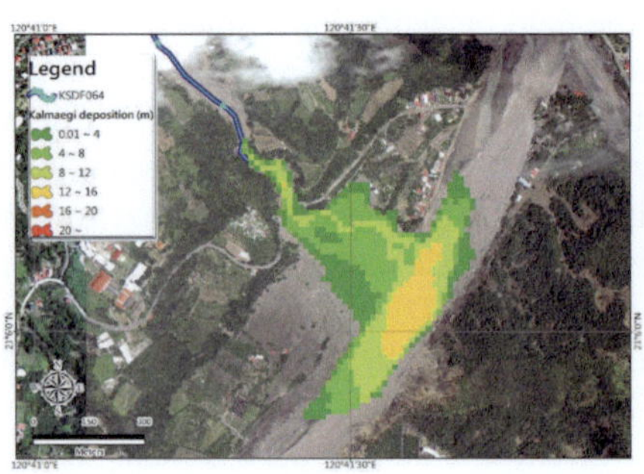

(a) (b)

Fig. 12 Simulated debris deposition of KSDF064 torrent caused by: **a** Typhoon Kalmaegi, and **b** Typhoon Morakot (base aerial photograph was taken after Typhoon Morakot)

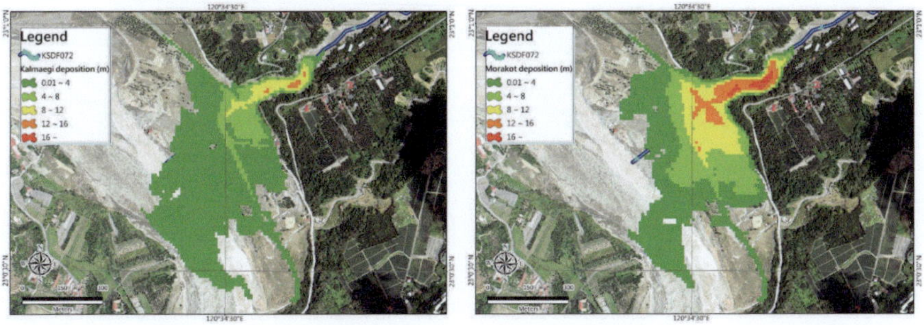

(a) (b)

Fig. 13 Simulated debris deposition of KSDF072 torrent caused by: **a** Typhoon Kalmaegi, and **b** Typhoon Morakot (base aerial photograph was taken after Typhoon Morakot)

than that in Typhoon Kalmaegi. The maximum simulated deposition depth was about 17.64 m, and the area with larger deposition depth was from torrent channel to the confluence with some deposition in mainstream Chishan River. The deposition at the confluence damaged the Huoshan Bridge, and the simulated results were quite consistent with the field investigation.

KSDF073 Basin. The inundated area and deposition depth of KSDF073 debris flow torrent from simulation were as shown in Fig. 14. Figure 14a shows the simulation results of Typhoon Kalmaegi, in which the debris deposited from the overflow point and concentrated along the KSDF073 torrent channel with the maximum deposition depth at the mainstream Launong River. Due to the overflow occurred in mid-channel, the deposition distributed along and was confined by the valley terrain till reaching the mainstream Launong River. The maximum simulated deposition depth was about 32.25 m at the mainstream Launong River. In field investigation, the debris filled the KSDF073 torrent channel near the confluence, which conformed with the numerical simulation well. Figure 14b shows the simulation results of Typhoon Morakot, and the inundated area was about the same as that in Typhoon Kalmaegi due to the mid-channel overflow and valley terrain. The maximum simulated deposition depth was about 34.2 m, and was larger than the 32.25 m in Typhoon Kalmaegi. The highest deposition depth occurred at the confluence and mainstream Launong River. The deposition at the confluence damaged the Chianshan No. 1 Bridge, and debris flowed into the mainstream Launong River reaching the opposite riverbank. The simulation results were quite consistent with the field investigation results.

5 Discussions

From the FLO-2D simulated results, the average deposition depth was obtained by averaging deposition depths over grids in the inundated area, and deposition volumes were also estimated. The inundated areas and average deposition depths caused by

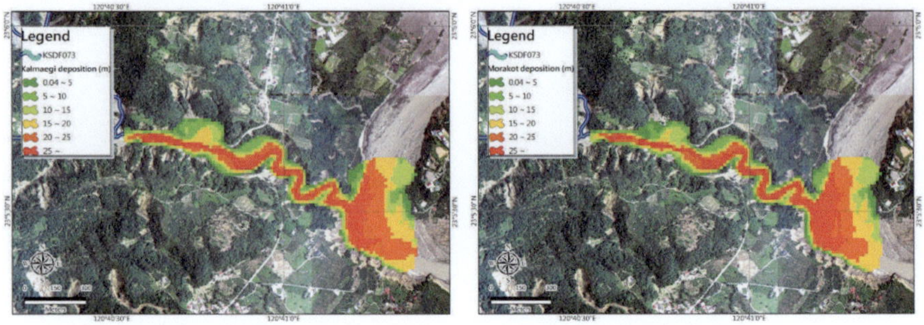

(a) (b)

Fig. 14 Simulated debris deposition of KSDF073 torrent caused by: **a** Typhoon Kalmaegi, and **b** Typhoon Morakot (base aerial photograph was taken after Typhoon Morakot)

Typhoon Kalmaegi and Typhoon Morakot events were listed in Table 2. A general trend of larger inundated area and average deposition depth caused by Typhoon Morakot than those by Typhoon Kalmaegi was found except for KSDF072 and KSDF073 for inundated area and KSDF073 for average deposition depth. From the field investigation and simulation results, debris overflowed mid-stream of KSDF072 and KSDF073 torrents, and were both confined by the valley terrain as shown in Figs. 13 and 14, especially for KSDF073. A large portion of debris flowed into the mainstream and flooded downstream channel of Launong River. Table 3 listed the simulated deposition volumes of Typhoon Kalmaegi and Typhoon Morakot, and field investigation results of Typhoon Morakot. The simulated deposition volumes in Typhoon Morakot were larger than those in Typhoon Kalmaegi except for KSDF073 torrent. Although the deposition volumes and average depths were almost the same for KSDF073 in the two events, the deposition distribution and the maximum deposition depth were different as illustrated in Fig. 14. It is suggested that the differences might be caused by the different rainfall patterns, and the similar conditions could be observed in all other four basins in Figs. 10, 11, 12 and 13. These differences in inundated area distribution, debris depth, and volume are important factors regarding mitigation. The simulated deposition volumes were almost one order larger than those estimated from field investigation in Typhoon Morakot. It might be due to that the debris deposition would have been removed or flooded away in the mainstream Launong River or Chishan River. Post event field investigations had difficulties in estimating debris volume removed or flooded away, however, evidences of debris deposition and flooding in mainstream were observed. Marchi et al. [8] suggested that debris volume in post event investigation could be underestimated due to debris been flooded by the mainstream especially in the case of narrow valley with small debris flow fans. Field investigations revealed evidences of huge volume of debris being carried downstream and flooded the downstream river channel. Most of the debris flow torrents in this study have narrow valleys and small debris fans. However, the differences between the simulated results and field investigation are quite large compared to those from other researches [1, 12].

Figure 15 shows the simulated deposition volume versus the accumulated rainfall of the five basins in both events, and it suggests a trend of higher accumulative rainfall leading to larger deposition volume except for KSDF073 torrent as discussed previously. The inundated area and deposition depth in Table 2 show a similar trend of increase with larger accumulated rainfall of Typhoon Morakot. No clear trend was found related to the rainfall intensity of the five basins in the two events. However, the different rainfall patterns could lead to different debris distribution condition as discussed previously. Figure 16 plotted the simulated deposition volume versus the inundated area. The range of data distribution conformed well with that by Iverson et al. [5], and the deposition volume appeared to increase with the inundated area in both events as well. A linear relation was fitted to the data of both events with good correlation. The slope is about 1/3 instead of 2/3 as proposed by Iverson et al. [5] with data from Lahar and debris flow. It is noted that both the deposition depth and inundated area increased with the deposition volume in Table 2 and Fig. 16. This suggests that not only the deposition volume is an important factor when evaluating hazard impact, a larger deposition volume also reflects that a larger inundated area and more severe impact due to higher deposition depth should be considered.

Table 2 Inundated areas and average deposition depths from simulation of Typhoon Kalmaegi and Typhoon Morakot

Torrent ID	Typhoon Kalmaegi		Typhoon Morakot	
	Inundated area, m^2	Deposition depth, m	Inundated area, m^2	Deposition depth, m
TNDF048	40,980	0.25	63,006	2.69
KSDF054	183,628	11.29	200,127	13.27
KSDF064	176,980	6.94	239,354	11.19
KSDF072	273,298	2.19	267,072	4.47
KSDF073	327,981	17.33	325,851	17.43

Table 3 Simulated deposition volumes and field investigation of Typhoon Kalmaegi and Typhoon Morakot (in m^3)

Torrent ID	Typhoon Kalmaegi	Typhoon Morakot	
	Simulated volume	Simulated volume	Field investigation
TNDF048	10,245	169,487	29,000
KSDF054	2,073,159	2,655,689	310,000
KSDF064	1,228,241	2,678,371	805,800
KSDF072	598,523	1,193,815	766,500
KSDF073	5,683,910	5,679,585	530,000

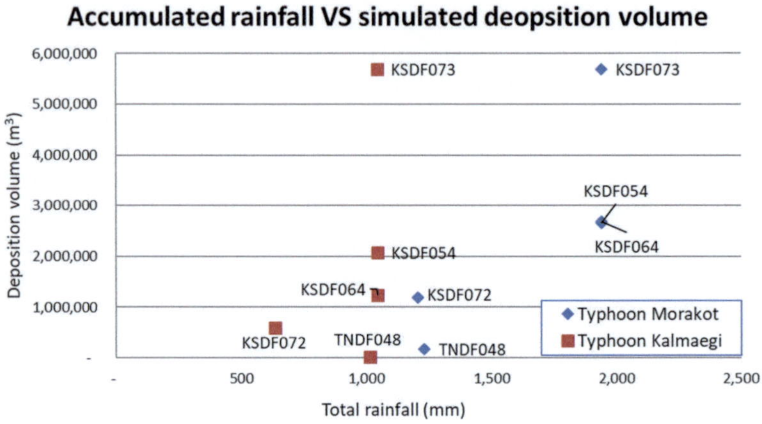

Fig. 15 Variations of simulated deposition volume versus accumulated rainfall of five debris flow basins

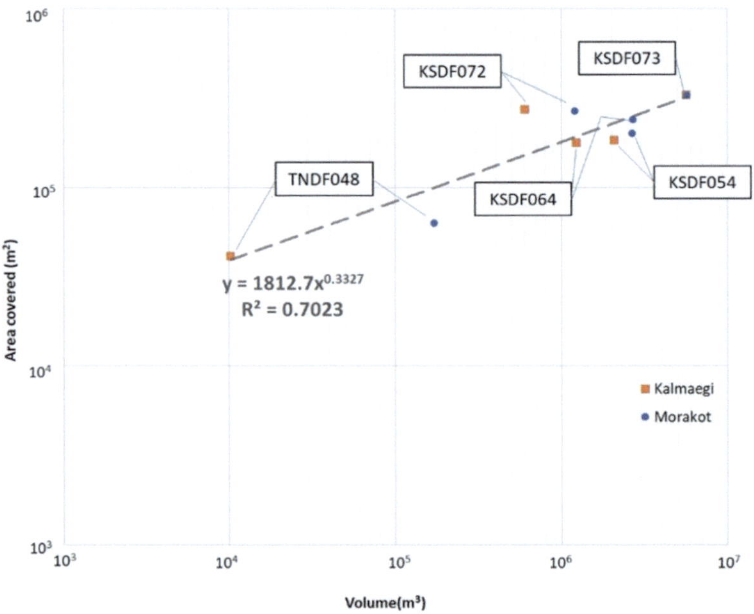

Fig. 16 Variations of simulated deposition volume versus the inundated area of the five debris flow basins in the two events

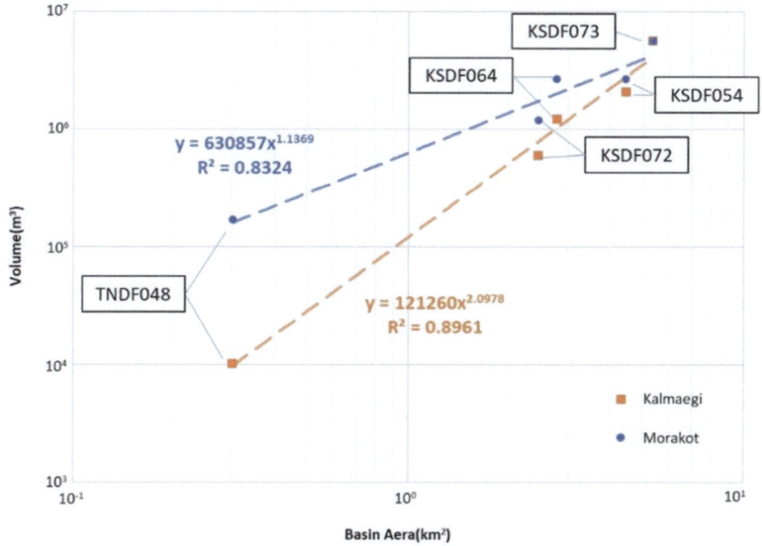

Fig. 17 Variations of simulated deposition volume versus basin area of the two events

Debris flow is attributed to the erosion and drainage of the basin, and the debris volume could be related to the basin area observing power law. The simulated deposition volume of the two events was plotted versus the basin area in Fig. 17. The higher accumulated rainfall of Typhoon Morakot caused larger deposition volume compared to that of Typhoon Kalmaegi as illustrated in the figure, and the differences in accumulated rainfall appeared to affect small basin more significantly and not as significantly as the basin area increased. It might suggest that the extreme rainfall has a more significant impact on smaller basin. The linear relation could be fitted to the two-event data sets separately as shown in Fig. 17. The simulated debris volumes in this study are much larger than the worldwide dataset and regression upper bound for northeastern Italy presented by Marchi et al. [8] as shown in Fig. 18. The slopes of the linear relation were about 1.137 and 2.098 for Typhoon Morakot and Typhoon Kalmaegi, respectively. The slopes of the lines are again higher than the relation listed by Marchi et al. [8]. This might suggest that in the extreme events such as Typhoon Morakot and Typhoon Kalmaegi, the data might not fit the general trend, but still appear to follow the linear relation. The two events occurred only one year apart, and both show the climate change tendency of high rainfall intensity and large accumulated rainfall but with different rainfall patterns. With such extreme events and extensive hazard, proper estimation of debris volume would require comprehensive consideration of the changes in precipitation characteristics caused by the climate change.

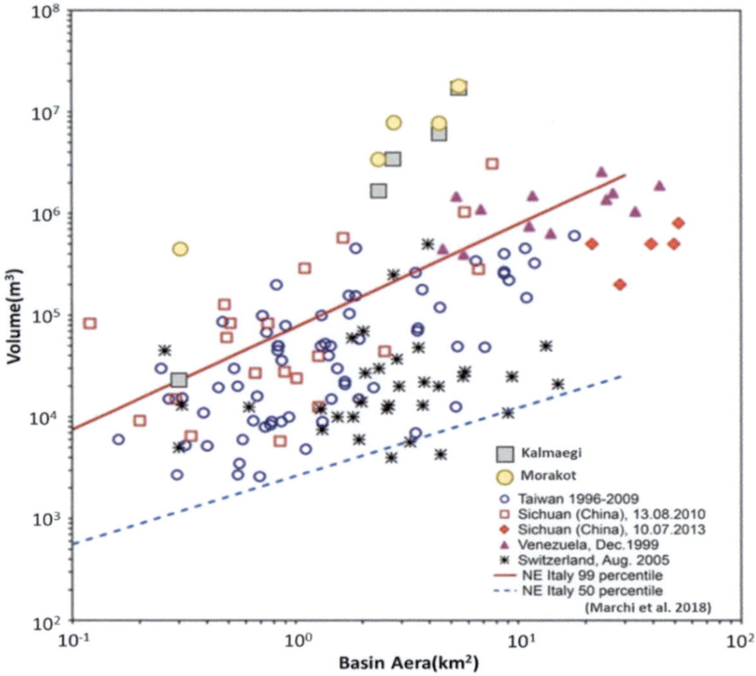

Fig. 18 Distribution of simulated debris volumes compared to the dataset by Marchi et al. [8]

6 Conclusion

Recurring debris flow triggered by the two extreme rainfall events, i.e., Typhoon Kalmaegi and Typhoon Morakot in five basins were investigated in this study. Both rainfall events were with high intensity and long duration but with very different patterns. Typhoon Kalmaegi brought high intensity but shorter duration rainfall, while Typhoon Morakot brought not so high intensity but a long duration with much larger accumulated rainfall. Field investigation and numerical simulation of debris flow caused by both events of five debris flow basins were conducted. A general trend of higher accumulated rainfall leading to higher debris deposition volume, larger affected areas, and deposition depth was observed, while no clear correlation was found related to the rainfall intensity. Such results suggested that the high accumulated rainfall caused by an extreme event could cause larger impacts of debris flow and lead to more severe damages. However, different rainfall patterns could induce different debris deposition distributions. A linear relation could be fitted to simulated deposition volume and basin area of the two events, and the simulated debris volumes are much larger than the worldwide dataset and regression upper bound for northeastern Italy presented by Marchi et al. [8]. The high accumulated rainfall of Typhoon Morakot causing large deposition volume appeared to affect small basin more significantly. It is suggested that the extreme events might not fit the general trend, but appear to follow the linear relation. Properly estimating debris volume related hazard would require more comprehensive consideration of the precipitation characteristics, which would be an important climate change issue need to be dealt with.

Acknowledgements Field investigation work in this research was supported by the Soil and Water Conservation Bureau, Taiwan (project number: SWCB-99-202, SWCB-101-004).

References

1. Brardinoni F, Church M, Simoni A, Macconi P (2012) Lithologic and glacially conditioned controls on regional debris-flow sediment dynamics. Geology 40(5):455–458
2. Chao CH (2004) Numerical simulation of debris flow and the characteristics of rheological parameters. Master Thesis, National Taiwan University, Taipei
3. Dore MHI (2005) Climate change and changes in global precipitation patterns: what do we know? Environ Int 31:1167–1181
4. Hürlimann M, Rickenmann D, Medina V, Bateman A (2008) Evaluation of approaches to calculate debris-flow parameters for hazard assessment. Eng Geol 102:152–163
5. Iverson RM, Schilling SP, Vallance JW (1998) Objective delineation of lahar-inundation hazard zones. Geol Soc Am Bull 110(8):972–984
6. Lin M-L, Wang K-L, Huang J-J (2005) Debris flow run off simulation and verification? case study of Chen-You-Lan watershed, Taiwan. Nat Hazards Earth Syst Sci 5(3):439–445
7. Lin M-L, Chen T-C, Hung F-I, Wang K-L, Lai T-L, Huang J-J, Chi W-L (2001) Potential analysis and hazard scenario simulation of debris flow rivers of Chen-You-Lan River watershed. Office of National Science & Technological Center for Hazards Mitigation, NAPHM, 90-08
8. Marchi L, Brunetti MT, Cavalli M, Crema S (2018) Debris-flow volumes in northeastern Italy: relationship with drainage area and size probability. Earth Surf Proc Land 44:933–943
9. O'Brien JS (2004) FLO-2D user's manual. Nutrioso, Arizona, June
10. Peng CH (2006) Debris flow simulation of central Taiwan area using FLO-2D. National Taiwan University, Master Thesis, Taipei
11. Rickenmann D (2016) Debris-flow hazard assessment and methods applied in engineering practice. Int J Erosion Control Eng 9(3):80–90
12. Santi PM (2014) Precision and accuracy in debris-flow volume measurement. Environ Eng Geosci 20(4):349–359
13. Sosio R, Crosta GB, Frattini P (2007) Field observations, rheological testing and numerical modelling of a debris-flow event. Earth Surf Proc Land 32(2):290–306
14. Trenberth KE (2008) The impact of climate change and variability on heavy precipitation, floods, and droughts. Encycl Hydrol Sci 17:1–11

Chain Disasters in Geotechnical Engineering Seen in the 2011 Great East Japan Disaster and Strategic Response

Motoki Kazama

Abstract There are several possible causes of a chain of natural disasters. In this study, chain disasters based on physical causal relationships and time-series chaining are classified. Subsequently, specific chain disaster examples that occurred in the Great East Japan Earthquake Disaster were introduced. Typical examples include the destruction of reservoirs and subsequent floods, liquefaction of tailings dams and subsequent mudflow disasters, environmental pollution, land subsidence, storm surge disasters, slope disasters, and river channel blockage. To what extent should such chain disasters be considered in the design? Following the earthquake, it served as an excuse to claim that the damage was unexpected. This excuse, however, cannot be used in future if such damage is identifiable. However, it is impractical to deal with all potential hazards, an amelioration strategy is required. An anti-catastrophic perspective is essential for the design of infrastructure and hardware.

Keywords Chain disasters · Great East Japan disaster · Geotechnical damage

1 Introduction

1.1 Great East Japan Disaster and the Reconstruction from the Disaster

The 2011 off the Pacific Coast of Tohoku Earthquake caused great damage to eastern Japan. This earthquake was named the Great East Japan Earthquake Disaster (GEJED). Owing to its magnitude (9–9.1 M_w), the damage was qualitatively and quantitatively higher than the previous magnitude 7 and 8 class earthquakes. Various geotechnical engineering issues were observed in the scene of damage and reconstruction [1, 2]. The government set a goal of ten years for reconstruction. As of

M. Kazama (✉)
Tohoku University, Aramaki, Aza-Aoba, Aoba-Ku, Sendai, Miyagi, Japan
e-mail: motoki.kazama.b2@tohoku.ac.jp

H. Hazarika et al. (eds.), *Multidisciplinary Approach for Disaster Resilience and Sustainable Technologies*, Lecture Notes in Civil Engineering 642,
https://doi.org/10.1007/978-981-96-8618-6_7

April 2022, reconstruction works other than the decommissioning of the nuclear power plant following the accident were almost completed.

1.2 Is a Triple Disaster a Complex Disaster?

It is well known that GEJED damages caused by the earthquake, tsunami, and nuclear power plant accident were dubbed a "triple disaster" in the newspapers immediately after the incident. In Japan's academic world, the term "complex disaster" began to be used. However, the definition of "complex disaster" is used with ambiguity. Complex disasters do not simply mean that different types of hazards occurred simultaneously, but that they are related to each other, resulting in increased damage. In general, complex disasters include not only natural disasters but also man-made disasters in human society (economic depression, panic, riots, war, etc.). However, this discourse will focus on the geotechnical disasters caused by natural disasters that overlap in time and space. Especially when planning and designing civil engineering facilities while considering the possibility of chain disasters.

In this paper, I will first consider the linkage of natural disasters by replacing "complex disasters" with "chain disasters". This change is necessary because, when considering the causality and linkage of disasters, the meaning of chain disasters is more precise than that of complex disasters. Next, I introduce the geotechnical chain disasters observed during the GEJED. Finally, we discuss how to deal with chain disasters.

2 What is Chain Disaster?

2.1 Classification of Chain Disaster Related to Geotechnical Engineering

First, there is the question of whether multiple events are interlinked in the physical mechanism responsible for their occurrence. It is also essential to understand how they overlap temporally and spatially. Temporal and spatial overlaps can be characterized by the mechanisms of disaster occurrence. For example, tsunamis following earthquakes have physical causality. In addition to the various system failures caused by antecedent events, it is also necessary to consider the scale of damage caused by secondary disasters caused by subsequent events (see Table 1).

When determining the level of fault tolerance for civil engineering facilities, the question is how much risk reduction is achieved. This is expressed as a scale of damage multiplied by the spatiotemporal probability of occurrence. Since most sizeable civil engineering facilities are public social infrastructure, the permissible

Table 1 Patterns of chain disasters due to the causality of complex events and relationships

	Causality and linkage of physical disaster events	
	Strong	Weak
Explanation	Antecedent events followed by subsequent events occurring in a chain within minutes, hours, or years	When independent events overlap temporally and spatially
Response	Chain event that should be considered: The response level requires social consensus according to the importance of the facility	When events are rare, simultaneity is rarely considered as a design condition

risk level is generally smaller than that used at private properties. The acceptable risk level requires social consensus depending on the importance of the facility.

Considering the abovementioned situation, the author classified chain disaster types into the following three cases with specific examples, as shown in Table 2.

(A) Directly causal cases in which antecedent events induce new subsequent severe events, so-called secondary hazards, should be considered to avoid catastrophic and fatal damage.
(B) Indirect causal cases in which antecedent events synergistically increase the damage of subsequent events: Cases in which antecedent events reduce or impair facility performance in responding to subsequent events. Consequently, it affects emergency recovery and relief activities immediately after the preceding event.
(C) Indirect weak causal case, in which the antecedent event increases the potential risk of a disaster from the subsequent event in the long term: low probability of temporal overlap between antecedent and subsequent events.

3 Chain Disaster Observed in the 2011 Great East Japan Disaster

3.1 Direct Causal Case, So-Called Secondary Disaster with Strong Causality

The Collapse of the Fujinuma Reservoir due to the Earthquake and Subsequent Flood Damage to the Downstream Area. The Fujinuma Reservoir, an agricultural reservoir in Sukagawa City, Fukushima Prefecture, collapsed because of the earthquake, and the flooded water turned into a mudslide downstream, causing a disaster. As a result, eight people died or went missing. According to an investigation report [3], the levee body slipped in seven stages due to the earthquake, and the initial slip toward the reservoir triggered subsequent overflow and erosion of the levee body,

Table 2 Specific examples of chain disasters related to geotechnical engineering

(A) Directly causal case in which the antecedent event induces a new subsequent event, so-called secondary disaster, with strong causality (see Sect. 3.1)
• River channel blockage due to slope failure
→ Floods downstream due to landslide dam collapse, debris flow disasters
• Collapse of the dam due to earthquake
→ Floods to downstream areas, debris flow disasters
• Earthquake-induced landslides, etc., flow into water areas
→ Occurrence of tsunami
Examples not related to geotechnical engineering:
• Damage to storage facilities for hazardous materials
→ Outflow of hazardous materials
• Building collapse due to seismic motion
→ Traffic disturbance, power loss, etc.
→ Large-scale fire due to loss of firefighting function
• Nuclear power plant accident
→ Radioactive contamination, etc.
→ Difficulty and delay in restoration work
(B) Indirect causal cases in which antecedent events synergistically increase the damage of subsequent events with medium causality (see Sect. 3.2)
• Settlement or destruction of seawalls due to seismic motion
→ Decrease in tsunami resistance performance
• Lifeline damage due to earthquake motion and tsunami
→ Increased damage due to obstacles to rescue and restoration activities
• Due to liquefaction of the ground, embankments, tide protection facilities, buildings' subsidence, and the foundation ground's softening
→ Deterioration of tsunami resistance, etc.
(C) Indirect weak causal case in which the antecedent event increases the potential risk of a disaster from the subsequent event in the long term with weak causality (see Sect. 3.3)
• Ground subsidence due to crustal movements caused by earthquakes
→ Vulnerability to flood damage
→ Obstacles to the gravity flow function of sewerage facilities
• Unstable sediment deposits due to volcanic ejecta and slope failures
→ Debris and mudflow disasters caused by heavy rainfalls and earthquakes
→ Promoting sedimentation in dam lakes and erosion control facilities
→ Decreasing disaster mitigation functions, etc.
• Drainage facility damage due to earthquakes
→ Delay in emergency restoration work
→ slope failure during heavy rainfall

causing the levee body to collapse. Given the severity of secondary disasters, it is critical to conduct seismic diagnosis and reinforcement of facilities that were constructed using the old standards [4]. In some areas, people came to live after reservoirs were built in areas that could be affected by secondary disasters from reservoir collapse (Photo 1).

Liquefaction of Tailings Dam and Subsequent Mudflow Failure. The level of the tailings dam located south of Kesennuma City in Miyagi Prefecture broke, and the liquefied slag flowed into the downstream river (see Fig. 1). Tailings are the residue after mineral extraction in a mine, it is a poorly graded fine silt with uniform

Photo 1 The right bank of the collapsed Fujinuma reservoir (by Prof. Seiki Kawagoe)

grain size and high liquefaction potential. Slag is often mixed with water and buried in sedimentation pits. Once liquefied, water does not separate from the soil particles, so it fluidizes and travels long distances. Such damage has been reported previously, including the 1978 North Izu earthquake in Japan and the 2010 Chile earthquake. Besides causing mudflows in downstream areas, there are also concerns regarding its heavy metal river contamination.

Other Related Cases Include Seismic Dam Formation Caused by Landslides. The 2008 Iwate-Miyagi Nairiku Earthquake [5], the 2008 Wenchuan Earthquake in China [6], the 2009 Typhoon Morakot that hit Taiwan, and the September 2011 Typhoon No. 12 in Japan are other examples of an increased risk of natural dam failure. The issues include countermeasures against secondary disasters, such as early detection of the formation of natural dams in mountainous areas; risk assessment; countermeasures; and evacuation measures if countermeasures cannot be implemented.

Fig. 1 Satellite image of liquefaction-induced fluidization and subsequent debris flow in a tailings dam, Kesennuma City, Miyagi prefecture (retouched Google earth image in 2011/4/7)

3.2 Indirect Causal Case, in Which Antecedent Events Synergistically Increase the Damage of Subsequent Events with Medium Causality

Decrease in Tsunami Resistance Performance of Pile Foundation Buildings due to Seismic Motion and Liquefaction. In Onagawa Town, Miyagi Prefecture, the tsunami destroyed middle-rise buildings constructed with reinforced concrete and steel. Among the collapsed buildings, there were spread foundations and pile foundations. However, the pile foundations may have suffered damage to the tops of the piles and reduced pull-out resistance because of shaking and liquefaction the tsunami arrived (see Photo 2) [7].

The Degradation of the Tsunami Resistance Seawall Facilities' Performance due to Ground Subsidence. In the 2011 off the Pacific coast of Tohoku earthquake, crustal deformation caused ground subsidence of several tens of centimeters around a wide area [8]. Because the crest height of the seawall decreased owing to ground subsidence before the tsunami hit, the tsunami resistance performance inevitably declined.

In addition, earthquake-induced ground subsidence is caused by the liquefaction of soft sandy soil layers and the consolidation settlement of soft cohesive soil layers [1, 9]. Of these, most subsidence due to liquefaction will end within a few minutes to an hour after the earthquake, leading to the deterioration of the wave

Fig. 3 A building with broken PC piles overturned because of the tsunami (by Prof. Mitsuji)

resistance performance of the breakwater facilities. During the 1995 Great Hanshin-Awaji Earthquake, the breakwater at Kobe Port subsided due to liquefaction, but no tsunami occurred; therefore, no complex disasters occurred.

The Possibility that the Damage Caused by the Shaking Reduced the Tsunami Resistance Performance. There were many cases where the soil of river embankments and road facilities was washed away by the tsunami. It is not clear how much damage was caused by the earthquake before the tsunami hit, but the earthquake damage may have exacerbated erosion and sediment washout caused by the tsunami. Photo 3 shows examples of soil washed away from the bridge attachment area and damage to sheet-pile quay walls. Damage from the tsunami may have expanded because the seismic motion damaged it. In the latter case, the water level in front of the quay dropped sharply as a result of the undertow, and the difference between the residual water pressure behind the quay increased. This is believed to be the reason for the loss of stability.

Liquefaction by the Mainshock and Re-liquefaction by Aftershocks. Once liquefied, the ground is likely to reliquefy. This is known as re-liquefaction phenomenon. If a large aftershock occurs before the excess pore water pressure in the ground liquefied by the mainshock dissipates, liquefaction may progress or continue during the aftershock. Twenty-nine minutes after the main shock, the Tokyo metropolitan area was greatly affected by an M7.7 aftershock with an epicenter off the coast of Ibaraki Prefecture [10]. In addition, even if the excess pore water pressure caused by the mainshock is completely dissipated, the soil particle structure may become looser than before the earthquake during the redeposition process after

Photo 3 Left: Mounting part of the bridge; right: damage to the sheet pile quay wall of Onahama port [provided by the Tohoku Regional Development Bureau (TRDB), Ministry of Land, Infrastructure, Transport and Tourism (MLIT)]

liquefaction. Photo 4 shows Onahama Port in Fukushima Prefecture, where large-scale liquefaction occurred due to the April 11th aftershock. This was also a type of chain disaster.

Photo 4 Onahama port (provided by the TRDB, MLIT))

3.3 Indirect Weak Causal Case, in Which the Antecedent Event Increases the Potential Risk of a Disaster from a Subsequent Event in the Long Term

Land Subsidence Caused by Earthquakes and Typhoon Storm Surge Disasters. Coastal lowlands, where seismic subsidence occurred, became significantly more vulnerable to storm surges and flood disasters. These lowland areas suffered a storm surge disaster in the August 2011 typhoon, after GEJED. Photo 5 shows a flooded area in Ishinomaki City, Miyagi Prefecture. Thus, the 1959 Ise-Bay Typhoon in Japan and Hurricane Katrina in the US are examples of land subsidence that caused severe storm surge disasters.

Consolidation Settlement of the Embankment Foundation Ground and Liquefaction of the Embankment Body. Typical damage to river embankments in inland areas has been noted. Figure 2 shows the results of an excavation survey of Naruse River's midstream and an outline of the mechanism. According to a survey, traces of liquefaction of the sandy soil embankment directly above the foundation ground composed of soft cohesive soil were observed [11]. The embankment moved laterally and broke into strips. In this area, traces of intrusion of sand veins toward the upper part of the embankment and traces of a thin (approximately 1 m at most) sand

Photo 5 State of Ishinomaki city two days after the earthquake and Tunami, flooded water in lowlands cannot escape (provided by the TRDB, MLIT)

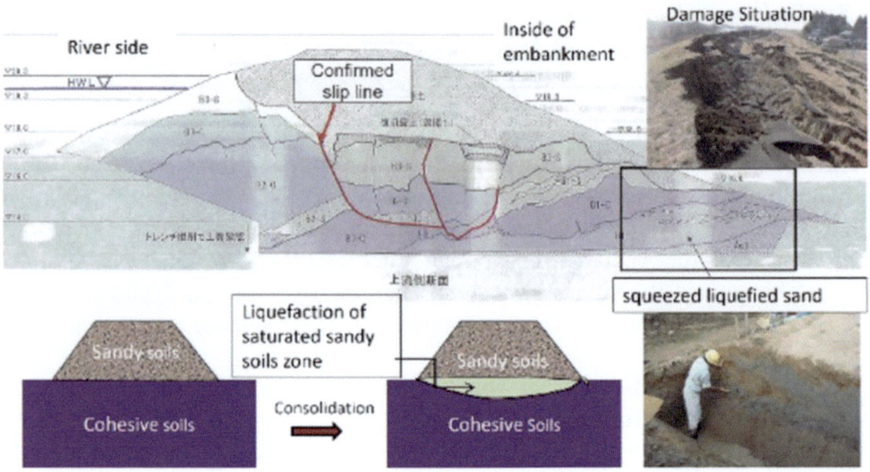

Fig. 2 Mechanism of typical damage seen in liquefaction damage to river embankments (liquefaction of embankment material) 30 km left bank of the middle stream of the Naruse River, excavation survey of the upstream area of Shimonakanome (by the TRDB, MLIT)

layer thought to have been liquefied were squeezed out to the side. It is thought that this is because the sandy soil, which is the embankment material, became saturated below the groundwater level due to the consolidation settlement of the foundation clayey layer and became loose owing to stress relaxation, so it was liquefied. It is a chain disaster because of liquefaction damage caused by the consolidation settlement of soft ground.

4 Response to Chain Disasters

4.1 Occurrence Patterns of Chain Disasters and Their Responses

So far, we have seen patterns of the occurrence of complex chain disasters. Since the frequency of occurrence of these disasters is low, there is currently no firm way of ascertaining, both technically and socially, the extent to which these disasters should be considered in facility design. In general, it is essential to take measures to avoid damage such as (A), which causes a large secondary disaster due to chain damage. On the other hand, damages such as (B) should be dealt with after carefully examining whether the chain and its complexities are fatal. However, cases like (B) introduced here have much room for reexamination since they have not been considered in the design so far. In the case of damage, such as (C), the occurrence probability of the event is independent, and there is time to respond.

4.2 The Word "Unexpected"

The word "unexpected" became popular in society during GEJED. Particularly in manufacturing, which is the subject of engineering, where certain design conditions are necessary when making things. The irresponsible stance that it is natural for a product to break if it exceeds its design conditions was socially criticized.

"Unexpected" refers to a quantitative excessive external force beyond what was assumed in the design, as described above. It also refers to design conditions that are not considered at all, such as the erosion resistance of the soil structure of the embankment at the time of overflow and the residual water pressure acting on the quay due to the tsunami backwash. In other words, failure modes that were not considered were unexpected. The latter can be considered a major reflection point.

In response to such a situation, "robustness," "ductility," and "redundancy" that does not fatally break even when an external force beyond consideration in design acts on the design object is required. This is an antithesis to the design concept that has sought to optimize and rationalize the design method of civil engineering facilities by advocating cost reduction. Providing robustness and tenacity to civil engineering facilities and giving them a margin in performance is a value that has a different direction from optimization and rationalization for simple cost reduction. Therefore, it is necessary to reform the consciousness of the design concept itself. Honda et al. [12] considered incorporating the concept of "anti-catastrophic" into its seismic design code, which requires resistance to external forces beyond those considered in the design. Kazama et al. considered "anti-catastrophic" strategy for liquefaction damage [13].

4.3 Safety Engineering Perspective for Civil Engineering Facilities

It is important to avoid fatal damage in the event of a disaster. In other engineering fields, such as machinery, electricity, and the manufacturing industry, there are many concepts of safety based on the premise of failure, as typified by keywords such as fail-safe, backup, passive safety, and damage-tolerant design. Civil engineering facilities are public facilities that benefit many citizens, have a much longer service life than home appliances and automobiles, and require a higher degree of safety. However, despite these characteristics and material deterioration, accidents are inevitable. There is no such thing as absolute safety and unbreakable facilities; based on this recognition, disaster resistance technology that incorporates ideas for how civil engineering facilities will break in the event of a disaster and ways to minimize damage after they break should be considered. This viewpoint should be a countermeasure from the viewpoint of setting the possible seismic ground motion level and tsunami wave height.

In addition, until now, seismic design has been used in a narrow sense to deal with the physical destruction of hardware facilities; however, the level of damage to hardware and the damage to system functions made up of hardware are different concepts. It is necessary to proceed with designing hardware facilities from the perspective of system malfunctions and business continuity planning (BCP).

4.4 Responding to Rare Mega-Disaster

GEJED is said to be a once-in-a-century disaster. According to the theory of extreme statistics, the probability that a 100-year-old civil engineering facility will encounter a once-in-100-year disaster at least once during its service period was 63.4%. Similarly, there was a 9.5% chance of encountering a once-in-a-thousand-year disaster. Unexpected external forces are commonly encountered. However, it is also true that there is enough time before the next event once the disaster has occurred.

After the earthquake, there were some non-scientific discussions because the actual experiences were traumatic in the areas affected by the tsunami. We should discuss the development of hardware for civil engineering facilities specializing in Level 2 tsunami disaster prevention. Similarly, there has been no scientific discussion on the safety of treating water containing low levels of radioactive materials.

5 Conclusion and Remarks

I hope this will provide readers with an opportunity to think about how to deal with rare natural disasters and unexpected events in terms of specific measures for recovery and reconstruction. I cannot help but think that facing a once-in-a-century disaster, a geological event is an arrogant act against nature when humans try to resist head-on.

Acknowledgements This work was supported by JSPS KAKENHI (grant number JP20H00258). The photos included in this text include those provided by Professor Kawagoe of Fukushima University, Professor Mitsuji of Yamagata University, and the Tohoku Regional Development Bureau of the Ministry of Land, Infrastructure, Transport and Tourism. I would like to express my most profound appreciation to my collaborators.

References

1. Kazama M, Kawai T, Kim J, Mori T (2015) The geotechnical issues of the damage caused by the great east Japan disaster and reconstruction for the Tohoku region. In: 15th Asian regional conference on soil mechanics and geotechnical engineering, ATC1-3-02

2. Noda T, Kazama M, Asaoka A (2017) New developments related to clarification of the mechanisms of ground deformation caused by earthquakes—in the wake of the great east Japan earthquake. JSCE 5:133–144. Special topic—restoration and recovery from the 2011 Great East Japan earthquake (invited paper)
3. Fukushima prefecture agricultural dam and reservoir earthquake resistance verification committee: investigation report (2012)
4. JGS (2012) Issues and countermeasures for geological disasters during earthquakes, Lessons learned and recommendations from the 2011 GEJED, Second Recommendation 6.8
5. Kazama M, Kataoka S, Uzuoka R (2012) Volcanic mountain area disaster caused by the Iwate-Miyagi Nairiku earthquake of 2008, Japan. Soils Founds 52(1):203–219
6. Qiang X, Fan X-M, Huan R-Q, Van Westen C (2009) Landslide dams triggered by the Wenchuan earthquake, Sichuan Province, south west China. Bull Eng Geol Environ 68(3):373–386
7. Sugimura Y, Mitsuji K (2012) Foundations of collapsed buildings, Part 2, proposals for future tsunami countermeasures. In: Architectural Institute of Japan symposium "Lessons from the Great East Japan earthquake, building a new country in the future"
8. Imakiire T, Koarai M (2012) Wide-area land subsidence caused by "the 2011 Off the Pacific Coast of Tohoku earthquake". Soils Foundations 52(5):842–855
9. Yasuhara, Kazama M (2015) Land subsidence of clay deposits after the Tohoku-Pacific ocean earthquake. In: Proceedings of IAHS, pp 211–216
10. Yasuda S, Harada K (2011) Liquefaction-induced damage in the reclaimed lands along Tokyo Bay. JGS 59(7):38–41
11. Kazama M, Kawai T, Mori T, Kim J, Yamazaki T (2015) Subjects of the liquefaction research seen to the liquefaction damage of the Great East Japan earthquake disaster. JAEE 15(7):49–59
12. Honda R, Akiyama M, Kataoka S, Takahashi Y, Nozu A, Murono T (2016) Seismic design method to consider "anti-catastrophe" concept. JSCE 72(4):I-459–472
13. Kazama M, Kamura A, Kim J, Han J (2020) Liquefaction research to anti-catastrophic strategy for external forces exceeding the design seismic level. In: 17th World conference of earthquake engineering, Sendai, Japan, Paper Number Cos-001

A Three stage Method for Debris Flow Analysis

Nicola Moraci, Maria Clorinda Mandaglio, and Mariantonietta Ciurleo

Abstract Rainfall-induced debris flows are characterised by high velocity, can be triggered by a slide and occur on established paths, usually first- or second-order drainage channels. During flow path, these phenomena entrain a strong quantity of material and water. Debris flow risk assessment and risk mitigation measure design involve forecasting spatial (susceptibility) and temporal (hazard) landslide distribution. The landslide susceptibility analysis provides the most susceptible areas to landslides and suggests where disaster protection measures are necessary to reduce the exposure of elements at risk and increase the community resilience. In this context, the paper proposes a methodology for debris flow susceptibility and hazard analysis. The methodology examines both debris flow triggering and propagation stages through two physically based models. The "Transient Rainfall Infiltration and Grid-Based Regional Slope-Stability" (TRIGRS) model provided information about the location and initial volume of potential landslides. The "Smoothed Particle Hydrodynamics" (SPH) model, using the triggering volumes as input data, allowed for back analysis of the propagation in terms of both main pathway and depositional area. The models can be easily implemented over large areas for fast landslide risk analysis and are able to provide interesting information for the design of disaster mitigation structures.

Keywords Fast landslides · Spatial landslide predictions · Analysis procedure

N. Moraci (✉)
University of Reggio Calabria, Reggio Calabria, Italy
e-mail: nicola.moraci@unirc.it

M. C. Mandaglio
University of Salerno, Fisciano, SA, Italy

M. Ciurleo
National Research Council of Italy, Research Institute for Geo-Hydrological Protection (CNR—IRPI), Rende, CS, Italy

H. Hazarika et al. (eds.), *Multidisciplinary Approach for Disaster Resilience and Sustainable Technologies*, Lecture Notes in Civil Engineering 642,
https://doi.org/10.1007/978-981-96-8618-6_8

1 Introduction

Rainfall-induced debris flows can be triggered by a slide and periodically occur on gullies and first- or second-order drainage channels [1]. They can reach velocities higher than 5 m/s, and are characterised by a strong entrainment of material and water along the propagation path.

Debris flows are often responsible for loss of human life and socio-economic consequences. Owing to the consequences related to debris flow occurrence, landslide forecasting (spatial—susceptibility—and temporal—hazard) is mandatory in landslide risk assessment and identifies the areas where risk mitigation measures should be implemented to reduce the exposure of element at risk.

Landslide susceptibility and hazard analysis can be pursued by different methods classified by Soeters and van Westen [2] in heuristic, statistical, and deterministic methods. Especially when deterministic methods are used, at the centre of the susceptibility and hazard assessment as well as the protection measure design there is the analysed landslide and the involved soils. Furthermore, when deterministic methods are applied over large areas, it is necessary to consider models able to simulate landslide triggering and propagation along three dimensions. Many studies are available on the analysis of landslide inception (e.g., Sorbino et al. [3], Schilirò et al. [4], Lizárraga and Buscarnera [5]) and propagation (e.g., Pastor et al. [6, 7], Pirulli and Sorbino [8]). On the contrary, there are no scientific papers which analyse both landslide inception and propagation over large area by one numerical model.

To this regard, the innovative nature of this work is to propose a methodology for debris flow susceptibility and hazard analysis over large area based on combining the "Transient Rainfall Infiltration and Grid-Based Regional Slope-Stability" (TRIGRS) model and the "Smoothed Particle Hydrodynamics" (SPH) model. In particular, TRIGRS is used to analyse the inception phase and SPH is used for analysing the propagation phase. The proposed methodology is calibrated and validated on two debris flows occurred in Calabria (southern Italy).

2 Materials and Methods

The area where the methodology was calibrated and validated is located in the western-south border of the Calabria Region (Fig. 1a). It extends approximately 1 km^2, it is bounded by a flat surface at the top and by a densely urbanised village at the bottom (Fig. 1b).

This area is periodically affected by shallow landslides as testified by several phenomena recorded during the years (Fig. 1b). Among these, debris flows occurred in 2001 and 2005 caused several damages at the bottom of the slopes, where Favazzina hamlet and several lifelines are located. Particularly, a debris flow initiated by two translational landslide triggering areas A_{01} and A_{02} (Fig. 1b) occurred on 12 May

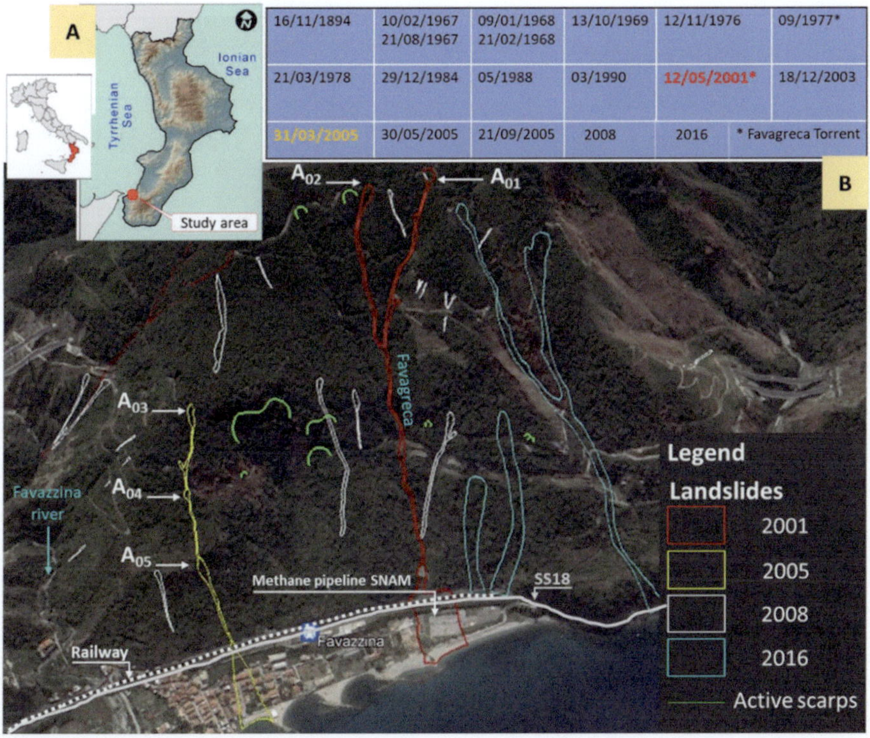

Fig. 1 The study area. **a** Location of the study area; **b** landslides date from 1894 to 2016 and 2001 and 2005 debris flows

2001 and during the propagation it hit the methane pipeline SNAM, the state road and caused the derailment of a train [9] (Fig. 1b).

Another debris flow was activated along the slopes located behind the town of Favazzina on 31 March 2005. It was characterised by three translational shallow slides (Fig. 1b) at about 370 m (triggering area A_{03}), 242 m (triggering area A_{04}) and 170 m a.s.l. (triggering area A_{05}). This debris flow hit the state road SS18, the Favazzina hamlet and caused the derailment of the ICN Reggio Calabria-Milan intercity train.

Owing the periodicity with which these phenomena occur on these slopes, it becomes essential to carry out susceptibility and hazard analyses. The methodology proposed to pursue this aim is shown in Fig. 2. The centre of the methodology is the landslide event that should be speedily identified in terms of type, geometry and rainfall data.

The proposed methodology consists of three stages necessary to study the whole landslide process from the inception to the deposition (Fig. 2). Stage I consists of database creation where rainfall, topographical and geomorphological data; mechanical and hydraulic properties of soils, pore water pressure regime and rheological

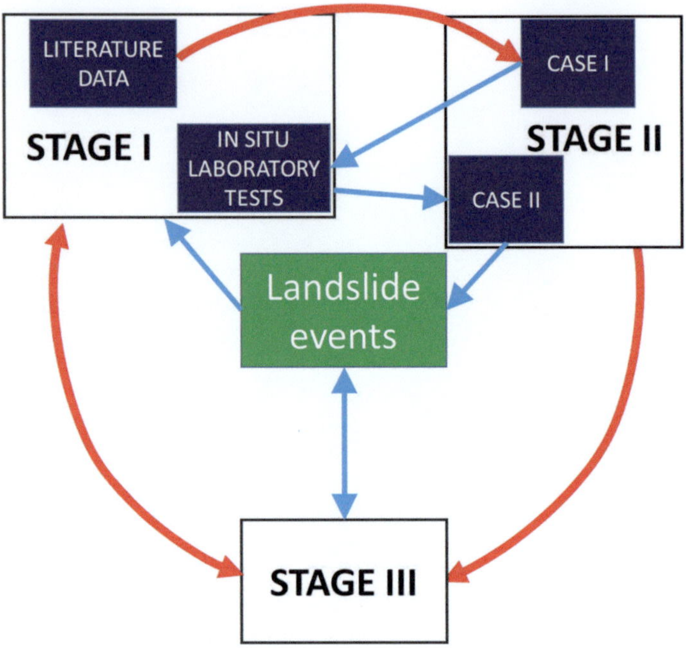

Fig. 2 The proposed methodology

model of the soil–water mixture are collected. The data base is a dynamic data base since it is created starting from the data available from scientific literature on soils similar to those studied but it can be upgraded every time a new data is available. Referring to geotechnical data, for dynamic and upgradable data base it means that when in-situ investigations and laboratory tests will be carried out, these go to populate the database (Fig. 2).

Debris flow inception areas are back-analysed by the "Transient Rainfall Infiltration and Grid-Based Regional Slope-Stability" (TRIGRS) model that couples an infiltration model [10, 11] with an infinite slope-stability model (Taylor, 1948) in stage II. This model was selected because of it is able to simulate these phenomena in the Favazzina study area [12–14]. TRIGRS allows us to distinguish unstable ($FS \leq 1$) from stable cells ($FS > 1$) in output.

It was used as a method to locate in-situ investigation and laboratory tests (Case I) and to produce a shallow landslide susceptibility map (Case II). In both cases, several parametric analyses were carried out using different input parameters among those of created database. Particularly in Case I, TRIGRS was used with geotechnical values available in the literature referred to the geographical context of the study area.

The best landslide susceptibility map, obtained by literature data, was chosen by typical statistical indicators [12] and on the slopes identified by Case I as the most susceptible to debris flows, further in-situ investigations that included sampling, and then geotechnical laboratory tests, were located on.

In Case II, TRIGRS was used as tool to obtain a shallow landslide susceptibility map using as input data the results of laboratory tests performed on soils involved in the surveyed landslides. This allowed to obtain an accurate shallow landslide susceptibility map from which triggering areas and mobilised initial volumes were extracted. The obtained results were used as input data for analysing the propagation and the deposition steps (Stage III). Stage III used the “Smoothed Particle Hydrodynamics” (SPH) model [6] with the following input data: triggering volumes, rheological parameters, and Digital Elevation Model (DEM).

The back analysis was verified using two dimensionless indices ($I_{\text{trig}}, I_{\text{dep}}$) defined for triggering (Stage II) and depositional analyses (Stage III), as follows:

$$I_{\text{trig}} = \frac{A_{\text{UTL}}}{A_{\text{TL}}} \times 100 \quad (1)$$

$$I_{\text{dep}} = \frac{A_{\text{SDF}}}{A_{\text{TDF}}} \times 100 \quad (2)$$

where A_{TL} are the in-situ observed source areas, A_{UTL} are the areas computed as unstable located within the A_{TL}, A_{TDF} is the in-situ observed debris fan and A_{SDF} is the numerically computed debris fan located within the A_{TDF}.

3 Results

Referring to the geometry of the two analysed landslides, classified as debris flows, the only available information is referred to the event occurred on 12 May 2001. According Bonavina et al. [15], this phenomenon presents triggering areas with a prismatic geometry and a slip surface located at 1.5 m of depth.

Regarding rainfall data, the available information is the daily precipitation of the Scilla rain gauge located near the analysed slopes. The rainfall data were, respectively, equal to 20 mm over two consecutive days on 12 May 2001 and 13.6 mm on 31 March 2005.

3.1 Stage I

Topographical and Geological data

Referring to topographical data, Digital Elevation Model (DEM) and Digital Surface Model (DSM) with pixel resolution of 5 m, 2 m and 1 m were available.

It is worth nothing that passing from DEM at 1:5000 scale to 1:1000 scale, several differences in terms of topographical and geometrical characteristics of the slope can be noticed (Fig. 3a, b). These differences become much more evident when passing from DEM to DSM (Fig. 3b, c) where the exposed elements appear.

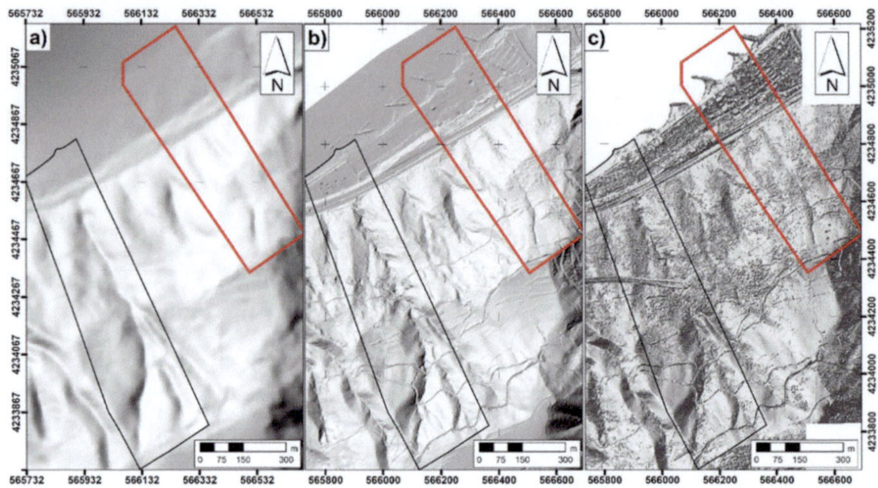

Fig. 3 Topographical data used in the analyses. **a** DEM (5 × 5 m); **b** DEM (1 × 1 m); **c** DSM (1 × 1 m). Black box borders the area involved by 2001 debris flow and red box borders the area involved by 2005 debris flow

From a geological point of view, the area is manly made up of high-grade metamorphic rocks (para- and ortho-gneiss), overlapped by sedimentary deposits [9, 16]. The outcropping rocks are strongly tectonised and deeply weathered gneiss, and they can be grouped in six weathered classes. Particularly, from the bottom to the top, the weathering sequence made up of Class II (slightly weathered rock) and Class III (moderately weathered rock) hardly outcropping, while Class IV (highly weathered rock) crops out in the middle-lower portions of the slope; Class V (completely weathered rock, or saprolite) prevails in the upper portion of the slope and in the top of the sequence Class VI (residual, colluvial and detrital soils) covers about 60% of the study area (Fig. 4).

Geotechnical and rheological data

In order to characterise the soil from a geotechnical point of view, first, the geotechnical data available in literature were summarised by Ciurleo et al. [13].

Both debris flows had a slip surface that involved gneiss of class VI and was located at 1.5 m. According to USCS, these soils were classified as silty sand (SM) and inorganic silt of medium compressibility with sand (ML) whose main physical properties are summarised in Table 1.

Referring to mechanical properties, gneiss of class VI show values of cohesion (c') going from 0 to 5 kPa and values of shear strength angle (φ') varying from 30° to 44° according to several research [4, 17].

Regarding the hydraulic properties of gneiss of class VI, according to Cascini et al. [18] and Schilirò et al. [4] (Schilirò et al. 2016), the values of saturated conductivity

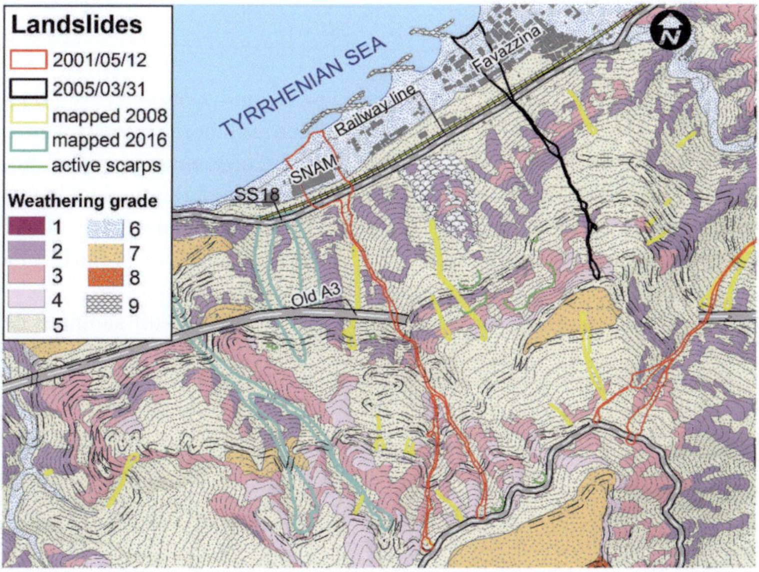

Fig. 4 Weathering-grade map of the study area and shallow landslide events (from Borrelli et al. [9] modified). Weathering-grade legend: (1) class II gneiss; (2) class III gneiss; (3) class IV gneiss; (4) class V gneiss; (5) class VI gneiss; (6) coastal and alluvial deposits; (7) marine deposits; (8) marine coarse deposits; (9) landslide debris

Table 1 Physical properties of soils of Class VI according to scientific literature

γ (kN/m^3)	γ_{sat} (kN/m^3)	γ_d (kN/m^3)	e (–)	n (–)	S (%)
15–20	19–22	12.5–16	0.65–1.15	0.40–0.55	43–99

(Ksat) range from 1.27E−06 m/s to 6.60E−05 m/s, and saturated volumetric water content θ_s varies from 0.38 to 0.4.

Referring to rheological data, the information available from scientific literature shows that the Bingham rheological model realistically simulates the soil water mixture behaviour of the debris flow that occurred in the Favazzina area [19].

The equation of the Bingham rheological model is reported as follows:

$$\tau = \tau_0 + \mu \frac{du}{dy} \tag{3}$$

where τ_0 (Pa) is the yield stress and μ is the dynamic viscosity (Pa s).

Yield stress and dynamic viscosity follow the exponential laws reported as follows:

$$\tau_0 = \alpha_1 \cdot e^{\beta_1 C_v} \tag{4}$$

$$\mu = \alpha_2 \cdot e^{\beta_2 C_v} \tag{5}$$

These laws depend on two empirical coefficients α_i and β_i obtained by a regression analysis on experimental data. To this regard, numerous laboratory tests were carried out by means of a rotary viscometer varying the solid concentration in volume from 20 to 45%. The results were interpreted using the Bingham rheological law.

For the majority of experimental data, the regression analysis of Eqs. 4 and 5 has provided the following values: $\alpha_1 = 0.251$ Pa and $\beta_1 = 0.132$; $\alpha_2 = 0.0112$ Pa s and $\beta_2 = 0.163$ (Fig. 5a).

Only some data, referring to a mixture of water and the soil sampled near the 2005 debris flow, follow different τ_0 and μ regression curves characterised by $\alpha_1 = 0.0002$ Pa and $\beta_1 = 0.27$; $\alpha_2 = 0.001$ Pa s and $\beta_2 = 0.21$ (Fig. 5b).

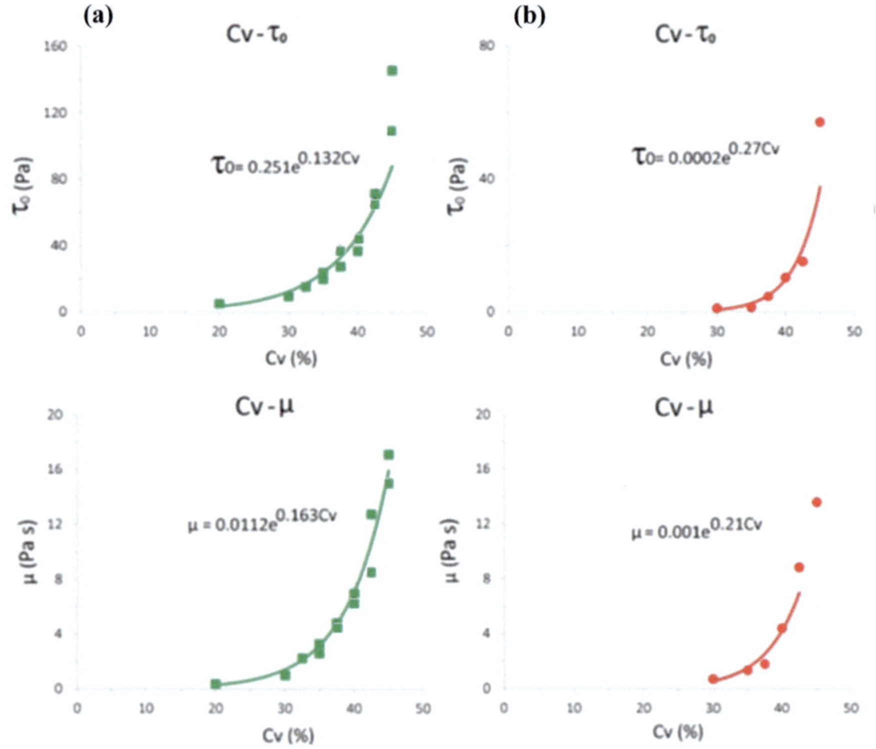

Fig. 5 The regression laws governing yield stress τ_0 and dynamic viscosity μ versus solid concentration in volume C_v (modified by Ciurleo et al. [20])

3.2 *Stage II*

Case I

In Stage II-Case I, preliminary TRIGRS analyses were firstly performed in the study area in order to properly located in-situ investigations and sampling. The input data used by TRIGRS are: rainfall data; DEM (5 m × 5 m); slope gradient and flow direction; the weathering-grade map; geotechnical data; initial pore water pressure regime.

Regarding rainfall data, the cumulative value recorded by the rain gauge located in Scilla in the two days preceding the event occurred on 12 May 2001 equal to 20 mm was used because it is larger than the similar one associated with the event occurred on 31 March 2005.

Regarding topographic data, a digital elevation model (DEM) with 5 m cell resolution was used. Slope gradient and flow direction were derived by the DEM.

Referring to weathering-grade map (Fig. 4), it is worth highlighting that the map was divided into two main lithologies: (i) gneiss of class VI (susceptible to shallow landslides of slide type) and (ii) other weathering classes of gneiss grouped as not interested by these phenomena.

According to the Unified Soil Classification System (USCS), and following the data available by scientific literature, the soils interested by shallow landslides can be classified as silty sand (SM, 27.26% of gravel, 50.58% of sand, 19.05% of silt and 3.11% of clay) and as inorganic silt of medium compressibility with sand (ML, sand = 30%, silt = 45%, clay = 25%) with plastic index and liquid limit equal to 9.23% and 33.27%, respectively.

According to scientific literature data, the physical and mechanical properties of class VI gneiss are summarised in Table 2.

Regarding pore water pressure, the water table was assumed at the ground surface in the "A" zone (Fig. 6) and at 1.5 m of depth (contact between class VI and less weathered gneiss) in the remaining part of the area (Zone B—Fig. 6).

The best landslide susceptibility map, obtained by literature data (Fig. 7) and chosen by typical statistical indicators [12], was found considering the average values of hydraulic parameters (K_{sat} = 1.79E−05 m/s, saturated hydraulic diffusivity D_0 = 7.92E−05 m^2/s, and θ_{sat} = 0.4), a constant value of soil cover thickness equal to 1.5 m [13], a value of intercept cohesion equal to 2.5 kPa and a value of the shear strength angle equal to 30°.

Table 2 TRIGRS input data for preliminary analyses (Stage II-Case I)

γ (kN/m^3)	c' (kPa)	ϕ' (°)	Thickness (m)	K_{sat} (m/s)	D_0 (m^2/s)	θ_{sat} (–)
20	0–5	30–40	1.5	1.75e−05	7.92e−05	0.4

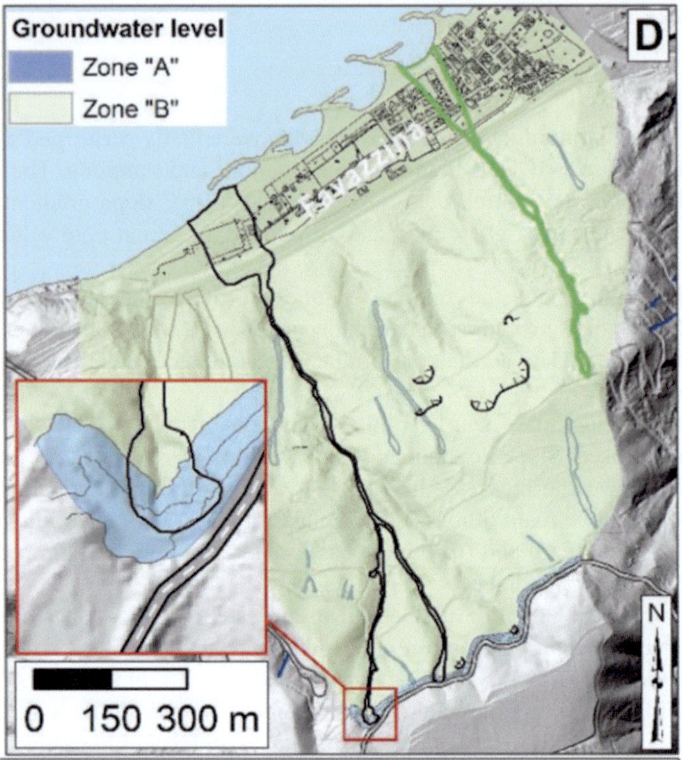

Fig. 6 Map of zones with given values of the initial depth of the groundwater table (from Ciurleo et al. [12] modified)

Starting from this base map, the in-situ investigations were located and performed in order to in-depth characterise the mechanical and hydraulic properties of weathered gneiss of class VI, especially in unsaturated conditions.

The in-situ investigations were located in the most susceptible slopes defined by Case I (Fig. 7a). They consisted of several seismic refraction tomographies and boreholes where Q5 undisturbed samples have been taken. Moreover, several undisturbed cubic were taken on areas where men and equipment were able to arrive (Fig. 7a).

Tomographies and boreholes allowed us to identify the main soil cover thickness (gneiss of class VI) that is potentially involved in shallow landslides. The gathered information shows that in the areas where shallow landslides occurred, the thickness value is about 1.5 m. Some examples were chosen in Fig. 7b.

Q5 undisturbed samples and the cubic ones were used to perform several laboratory tests in order to classify and mechanically and hydraulically characterise the residual, colluvial, and detrital soils (gneiss of class VI), including grain-size distribution, shear strength properties and soil permeability.

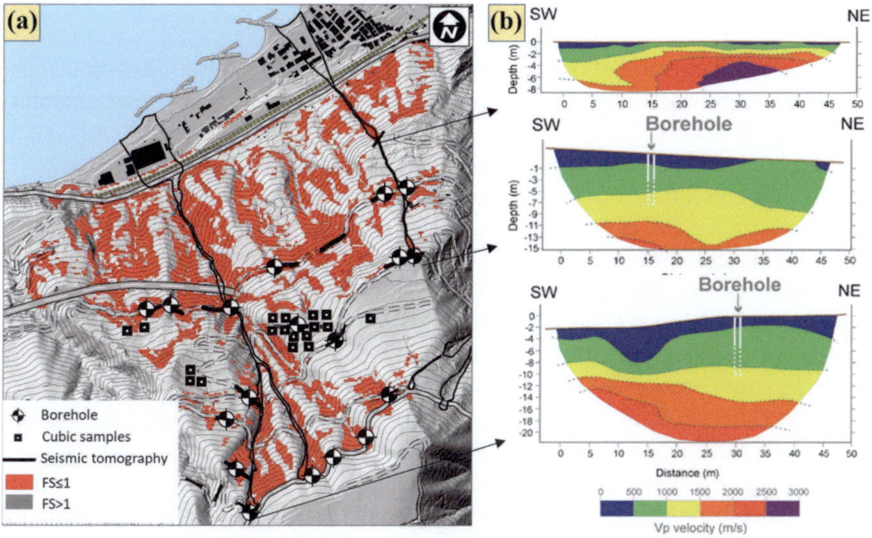

Fig. 7 **a**: Preliminary TRIGRS results: Stage II-Case I. **b** Seismic topography profiles in the source areas of 2001 and 2005 debris flows (from Ciurleo et al. [12] modified)

Based on grain-size distribution, the soils were classified as sands with clayey silts with uniformity coefficient (U) about equal to 76, 77, 113 with inorganic fine fraction, of medium compressibility and not active, the value of LL ranged from 30 to 33% and PI from 6 to 11%.

These soils have a unit weight (γ_s) varying from 25.9 to 26.1 kN/m^3; void index (e) variable from 0.9 to 1.17, and soil porosity (n) varying from 0.47 to 0.54.

Regarding hydraulic properties of soils, some permeability tests were performed in oedometer at different vertical stress respectively 50, 100, and 200 kPa, and the corresponding values of soil-saturated permeability ranged from 3.64E−07 to 1.63E−08 m/s.

The soil water characteristic curve was obtained by a suction controlled oedometer test, and the laboratory data were fitted by the Gardner model.

Referring to mechanical properties, the results of saturated triaxial tests carried out on undisturbed samples clearly show a ductile and contracting behaviour and the obtained shear strength parameters have been equal to $c' = 1$ kPa and $\varphi' = 35°$.

The geotechnical parameters of gneiss of class VI derived from laboratory tests were summarised in Table 3.

Table 3 Geotechnical data of class VI gneiss

c' (kPa)	φ' (°)	K_{sat} (m/s)	Θ_s (–)	Θ_r (–)	α (m^{-1})	Soil cover thickness (m)
1	35	3.64E-07	0.51	0.23	0.69	1.5

Case II

Stage II-Case II consists of performing several parametric analyses in saturated and unsaturated conditions using the data gathered by in-situ investigations and laboratory tests (Stage II-Case I).

The performed parametric analyses are scheduled in Table 4.

The first column of the table reports a progressive identification number with the letter U, indicative of unsaturated conditions, and S for saturated conditions. The following two columns report the value of cohesion and shear strength angle obtained by laboratory tests. The remaining columns report the value of saturated conductivity function, saturated and residual volumetric water contents and the Gardner's parameter (α). In all cases, the hydraulic value obtained by an oedometer test with a confining stress applied of 50 kPa has been considered.

The last three columns report the value of soil cover thickness, and the location of the water table in the two considered zones.

Referring to the soil cover thickness distribution over large area, the depth of 1.5 m was considered constant in the whole study area for the soil involved in the landslides, because the slip surface of shallow landslide was generally located at 1.5 m and the in-situ investigations confirmed a thickness of gneiss of class VI equal to 1.5 m in several parts of the slopes where shallow landslides occurred (Table 4).

Regarding initial pore water pressure regime, due to the lack of data, different positions of the initial water table were assumed in order to investigate the role of partially saturated conditions on the stability evaluation of the slopes (Table 4—cases from 1 to 3 U).

After that, saturated conditions were investigated. Firstly, a saturated analysis considering an initial steady state water table located at 1.5 m in the whole study area was performed (Table 4—case 4S).

Secondly, according to Ciurleo et al. [13], the study area was divided into two different zones, zones "A" and "B" (Fig. 6).

Zone "A" is a buffer zone made up of contour lines, each equal to 5 m which starts from a secondary road located near the 2001 debris flow triggering areas. Zone "B" is the remaining part of the study area (Fig. 6).

Table 4 TRIGRS input data for debris flow susceptibility analyses (Stage II-Case II)

CASE	c' (kPa)	φ' (°)	K_{sat} (m/s)	Θ_s (–)	Θ_r (–)	α (m^{-1})	Soil cover thickness (m)	Water table in "A" zone (m)	Water table in "B" zone (m)
1U	1	35	3.64E−07	0.51	0.23	0.69	1.5	3.0	3.0
2U	1	35	3.64E−07	0.51	0.23	0.69	1.5	2.5	2.5
3U	1	35	3.64E−07	0.51	0.23	0.69	1.5	2.0	2.0
4S	1	35	3.64E−07	0.51	0.23	0.69	1.5	1.5	1.5
5S	1	35	3.64E−07	0.51	0.23	0.69	1.5	0	1.5

This subdivision provided the opportunity to further investigate the influence of a secondary road on shallow landslide triggering using laboratory data that was known influenced the triggering of the 2001 debris flow [13, 17]. To this regards, the performed analyses considered a water table located at 1.5 m and 0 m from the ground surface in zone "A" (Fig. 6), and a steady state water table located at 1.5 m of depth, that is the bedrock-soil interface in the remaining study area (zone "B", Fig. 6) (Table 4).

The simulation performed in unsaturated conditions with an initial water table located at 3.0 m (Fig. 8 case 1U) shows that only some cells of the study area are computed as unstable by the model and this initial configuration did not reproduce the in-situ observed source areas available by the multi-temporal landslide inventory.

The results obtained when the initial water table is located at 2.5 m and 2.0 m from the ground surface (Fig. 8 cases 2U and 3U) show an increase of computed unstable cells even if the overall computed instability is still very far from the simulation of the triggering areas mapped by the multi-temporal landslide inventory.

In saturated conditions and with an initial water table located at the contact between residual, colluvial and detrital soils and less weathered rock, the obtained results (Fig. 8, case 4S) generally show a good fit between the unstable computed

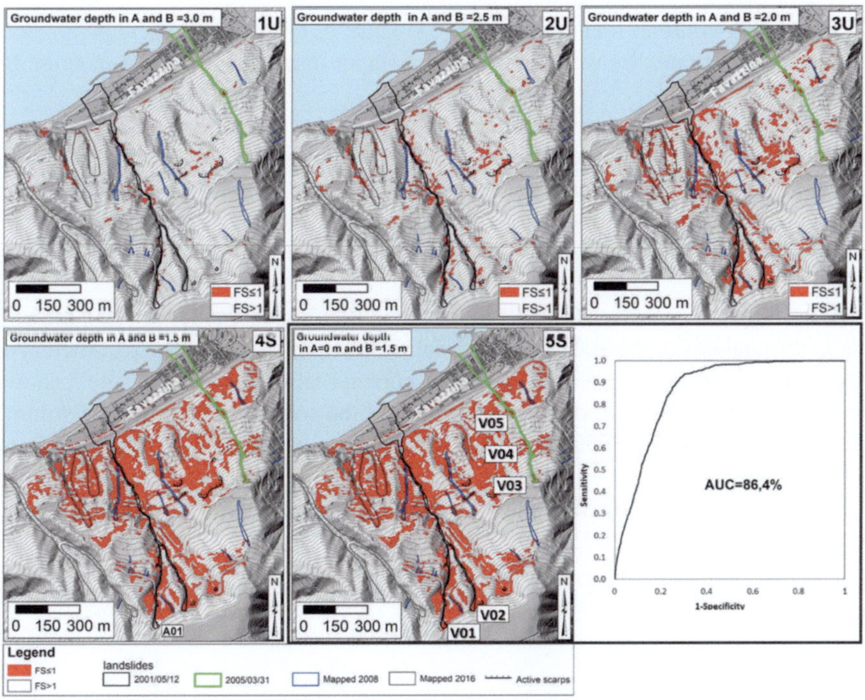

Fig. 8 Stage II-Case II results (from Ciurleo et al. [12] modified)

cells and the in-situ observed source areas expect for the triggering area A_{01} that is one of the source areas that was triggered the debris flow in 2001.

On the contrary, considering an initial water table located at 0 m from the ground surface in the zone "A" and at 1.5 m from the ground surface in the remaining part of the study area (Fig. 8, case 5S), TRIGRS is able to correctly back-analyse the landslide source areas mapped by the inventory including the triggering area A_{01} so providing a value of I_{trig} equal to about 90%.

For Case 5S, the area under the receiver-operating characteristic curve (AUC) was also computed to further justify the success of the analysis (Fig. 8). This value is equal to 86.4%.

Considering the best obtained simulation (Fig. 8 Case 5S), the computed triggering volumes (V_{01} and V_{02}) for 2001 debris flow are 900 m^3 and 1125 m^3 (triggering volumes 1 and 2 referred to A_{01} and A_{02}), for the 2005 debris flow, the computed triggering volumes (V_{03}, V_{04} and V_{05}) are 325 m^3, 187 $m^3$and 761 m^3 (Fig. 8 Case 5S).

3.3 *Stage III*

The triggering volumes identified in Stage II-Case II, DEM, rheological parameters and the "growth rate" Es were considered as input data of SPH model in stage III (Table 5).

An iterative analysis changing topographical data, rheological parameters and erosion coefficient was carried out.

Referring to topographical data, the analyses started from DEM at 1:5000 scale, to be coherent with the topographical data used by TRIGRS. Subsequently, several analyses were performed using DEM or DSM at different scales (1:1000, 1:2000, 1:5000).

According to Gioffrè et al. [16], the Bingham rheological model resulted the best one in the analysis of the 2001 and 2005 debris flow propagation. The parameters of this model (τ_0 and μ_b) were obtained by viscometer laboratory tests. Table 5 shows the value assumed by τ_0 and $\mathrm{\mu_b}$ as a function of the solid concentration by volume (C_v) ranging from 50 to 55%. The "growth rate" Es was considered variable from 0.001 to 0.002 m^{-1}.

The results obtained by SPH analyses were compared with propagation paths and debris fans of 2001 and 2005 debris flows.

Table 5 SPH input data

	C_v (%)	τ_0 (Pa)	μ_b (Pa·s)	Es (m^{-1})
2001 debris flow	50–55	125.56–477.14	37.42–115.84	0.001–0.002
2005 debris flow		184.51–356.99	38.79–87.63	

In order to analyse the role played by topographical data, value of C_v and Es are fixed constant ($C_v = 52\%$ and Es $= 0.0018\ \text{m}^{-1}$) and the parametric analyses initially contemplate the use of DEM or DSM at different scale. So doing, the best performance in terms of simulation of propagation path and depositional area of 2005 debris flow was obtained considering a DSM at 1:2000 scale. Indeed, the propagation path, that appears significantly larger than the real one at 1:5000 scale, was realistically reproduced at 1:2000 scale (Fig. 9a, b). At this scale, the value assumed by I_{dep} was equal to 78.42% that is the highest value reached fixing constant Cv ad Es and varying only topographical data (Fig. 9). Figure 9 also shows that a DSM at 1:1000 scale does not improve the value of I_{dep} and causes the divergence of some smoothed particles that go out to the main channel simulating a condition that is not coherent with the real phenomenon.

Focusing only on DSM at 1:2000 scale, the performed SPH analyses summarised in Fig. 9 show that the highest value of I_{dep} (92.35%) was obtained considering a solid concentration by volume equal to 50% and a value of growth rate equal to $0.0018\ \text{m}^{-1}$.

Regarding 2001 debris flow, the best performance of SPH model (I_{dep} equal to 72.0%) was obtained considering a DSM at 1:2000 scale, a value of $C_v = 54\%$ and Es $= 0.0015\ \text{m}^{-1}$ (Fig. 10).

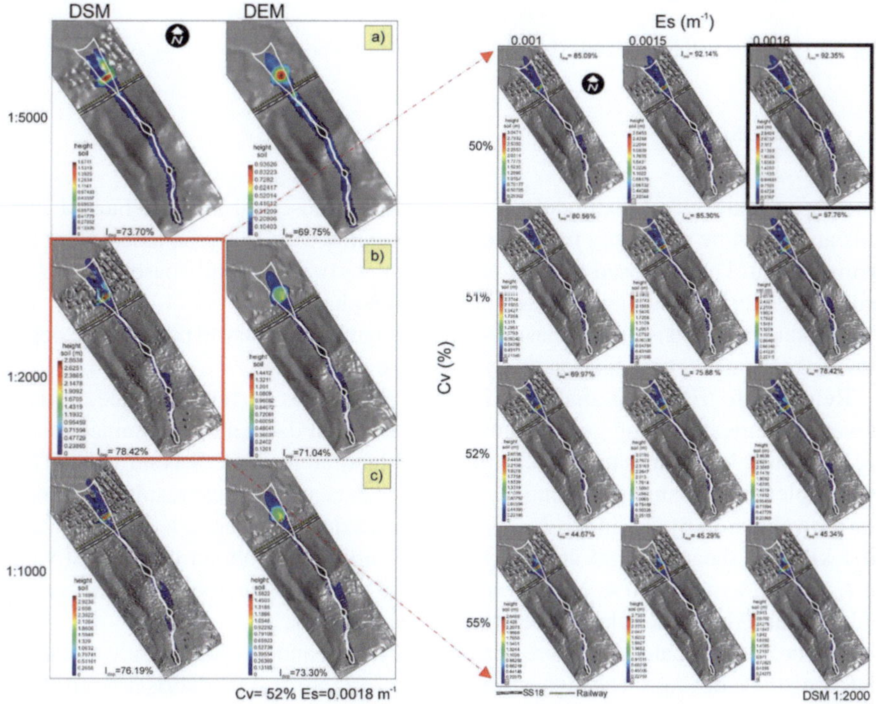

Fig. 9 2005 debris flow analyses by SPH (from Ciurleo et al. [20] modified)

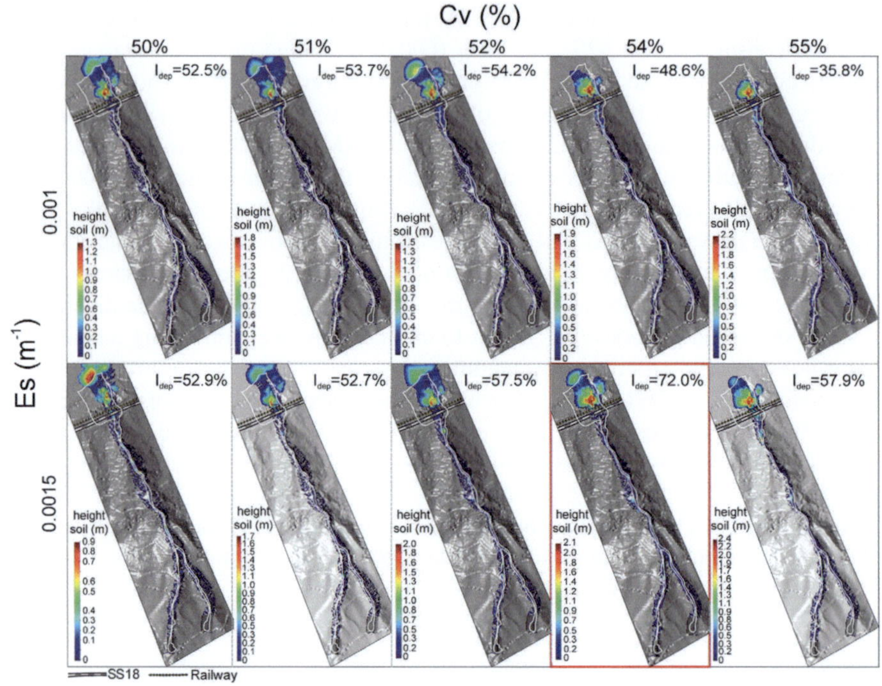

Fig. 10 SPH simulations for 2001 debris flow (from Ciurleo et al. [20])

4 Conclusion

The paper deals with a methodology for debris flow susceptibility and hazard analysis and it can be profitably used also for debris flow risk analysis, assessment and management. The proposed methodology combines two physically based models for analysing debris flow triggering and propagation phases.

The starting point is the database creation (Stage I) necessary to perform the debris flow triggering and propagation analyses (Stage II and III).

Referring to triggering analyses, TRIGRS model was performed twice, initially it was implemented using geotechnical literature input data and the obtained results were used to locate in-situ investigations and laboratory tests. Secondly, TRIGRS was implemented using geotechnical laboratory data and the best result was chosen for the validation by the index I_{trig}. For the Favazzina study area, a cohesion value of 1 kPa and shear strength angle equal to 35° allowed to obtain the highest value of I_{trig} equal to about 90%. Moreover, TRIGRS provided the debris flow triggering volumes used as input data for the propagation analyses.

The propagation analyses were performed by SPH and the validation was carried out using the index I_{dep}. The highest values of I_{dep} (72% for 2001 debris flow and 92%

for 2005 debris flow) were obtained considering $C_v = 54\%$ and Es = 0.0015 m^{-1} for 2001 debris flow and $C_v = 50\%$ and Es = 0.0018 m^{-1} for 2005 debris flow.

The good results obtained by applying the proposed methodology to the Favazzina case study showed a good performance of the models in the analysis of debris flow inception areas and debris flow main pathways so opening the way for a generalisation of the methodology over large area. Indeed, the two distributed physically based models can be easily implemented over large area on slopes potentially affected by debris flow with the aim to analyse debris flow susceptibility, hazard and risk. This allows, on the one hand, to use the results obtained by triggering analyses to locate debris flow monitoring systems on the areas potentially affected by these phenomena so improving both the probability to instrument slopes that will be affected by debris flows and the knowledge on debris flow predisposing factors.

On the other hand, once debris flow risk analysis was carried out, value judgments and risk tolerance criteria can be used to evaluate the risk and to define possible debris flow mitigation works to be implemented and designed. If debris flow mitigation measures are necessary, the most used ones are the protection structures that reduce debris flow runout and velocity, stop the phenomenon so protecting the element exposed to risk.

In the last years, the attention of several scientists focuses on the design of debris flow sheltering structure (deformable or rigid) that is based on the estimation of debris flow height and velocity and impact load [21].

Furthermore, for the best simulations of the 2005 and 2001 debris flows, the proposed methodology allowed to obtain the trend of debris flow height and velocity in several sections located along the propagation path. Figure 11 shows the results obtained at the centre of the O–O′ section located above a main road, that is the first exposed element hit by the flow during the propagation path (Fig. 11a, b).

From the Figure, it can be observed that in both cases debris flow velocity reaches a peak value higher than 20 m/s and an average value of about 5 m/s while debris flow height is about 1.5 m for 2001 debris flow and about 2.4 m for 2005 debris flow.

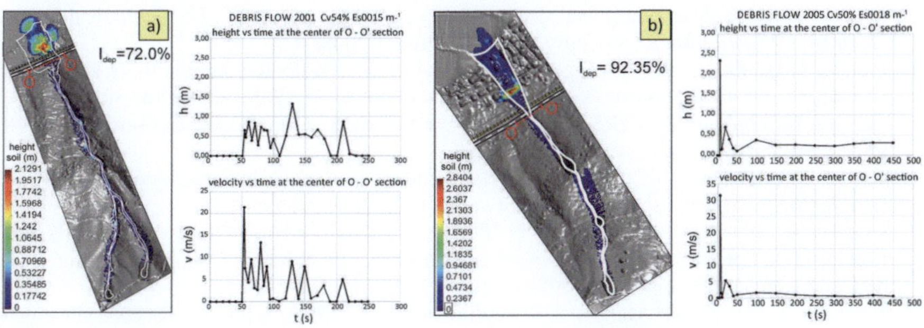

Fig. 11 The best obtained SPH simulations of the analysed debris flows with the respective variation of debris flow height and velocity in time at the centre of O–O′ section. **a** debris flow occurred in 2001; **b** debris flow occurred in 2005.

The highest values were reached by 2005 debris flow simulation (Fig. 11b). In this case, the highest value reached by peak velocity is about 30 m/s.

Acknowledgements All three authors contributed equally to the development of the research and the preparation of the manuscript.

References

1. Hungr O, Picarelli L, Leroueil S (2014) The Varnes classification of landslides-an update. Landslides 11:167–194
2. Soeters R, van Westen CJ (1996) Slope instability recognition, analysis and zonation. In: Turner AK, Schuster RL (eds) Landslides investigation and mitigation. TRB Special Report 247 National Academy Press, Washington D.C., pp 129–177
3. Sorbino G, Sica C, Cascini L (2010) Susceptibility analysis of shallow landslides source areas using physically based models. Nat Hazards 53:313–332
4. Schilirò L, Esposito C, Scarascia Mugnozza G (2015) Evaluation of shallow landslide triggering scenarios through a physically based approach: an example of application in the southern Messina area (northeastern Sicily, Italy). Nat Hazards Earth Syst Sci 15:2091–2109
5. Lizárraga JJ, Buscarnera G (2019) Spatially distributed modeling of rainfall-induced landslides in shallow layered slopes. Landslides 16:253–263. https://doi.org/10.1007/s10346-018-1088-8
6. Pastor M, Haddad B, Sorbino G, Cuomo S, Drempetic V (2009) A depth integrated coupled SPH model for flow-like landslides and related phenomena. Int J Num Anal Meth Geomech 33–2:143–172
7. Pastor M, Blanc T, Haddad B, Petrone S, Sanchez MM, Drempetic V, Issler D, Crosta GB, Cascini L, Sorbino G, Cuomo S (2014) Application of a SPH depth-integrated model to landslide run-out analysis. Landslides 11:793–812. https://doi.org/10.1007/s10346-014-0484-y
8. Pirulli M, Sorbino G (2008) Assessing potential debris flow runout: a comparison of two simulation models. Nat Hazards Earth Syst Sci 8:961–971
9. Borrelli L, Gioffrè D, Gullà G, Moraci N (2012) Suscettibilità alle frane superficiali veloci in terreni di alterazione: un possibile contributo della modellazione della propagazione. Rendiconti Online Società Geologica Italiana 21:534–536
10. Baum RL, Savage WZ, Godt JW (2008) TRIGRS-A Fortran program for transient rainfall infiltration and grid-based regional slope-stability analysis, version 2.0. US Geological Survey Open-File Report 2008–1159. http://pubs.usgs.gov/of/2008/1159
11. Iverson RM (2000) Landslide triggering by rain infiltration. Water Resour Res 36(7):1897–1910
12. Ciurleo M, Ferlisi S, Foresta V, Mandaglio MC, Moraci N (2022) Landslide susceptibility analysis by applying TRIGRS to a reliable geotechnical slope model. Geosciences 12:18. https://doi.org/10.3390/geosciences12010018
13. Ciurleo M, Mandaglio MC, Moraci N (2019) Landslide susceptibility assessment by TRIGRS in a frequently affected shallow instability area. Landslides 16:175–188
14. Ciurleo M, Mandaglio MC, Moraci N, Pitasi A (2020) A method to evaluate debris flow triggering and propagation by numerical analyses. Lecture Notes in Civil Engineering 2020, 40:33–41. https://doi.org/10.1007/978-3-030-21359-6_4
15. Bonavina M, Bozzano F, Martino S, Pellegrino A, Prestininzi A, Scandurra R (2005) Le colate di fango e detrito lungo il versante costiero tra Bagnara Calabra e Scilla (Reggio Calabria): Valutazioni di suscettibilità. G Geol Appl 2
16. Gioffrè D, Moraci N, Borrelli L, Gullà G (2016) Numerical code calibration for the back analysis of debris flow runout in southern Italy. In: Landslides and engineered slopes. experience, theory and practice: proceedings of the 12th international symposium on landslides (Napoli, Italy, 12–19 June 2016), vol 3, pp 1339–1344

17. Antronico L, Cotecchia F, Cotecchia V, Gabriele S, Gullà G, Iovine G, Lollino G, Moraci N, Pagliarulo R, Petrucci O, Rocca F, Sorriso-Valvo M, Terranova O (2006) Relazione finale Lotto 5—Attività di monitoraggio di siti in frana e di aree soggette a fenomeni di subsidenza. In: Azione 1.4.c—Azioni di studio, programmazione, sperimentazione, monitoraggio, valutazione e informazione finalizzati alla predisposizione e gestione di politiche integrate d'intervento di difesa del suolo. POR Calabria 2000–2006
18. Cascini L, Gullà G, Sorbino G (2006) Groundwater modelling of a weathered gneissic cover. Can Geotech J 43:1153–1166
19. Moraci N, Mandaglio MC, Gioffrè D, Pitasi A (2017) Debris flow susceptibility zoning: an approach applied to a study area. Rivista Italiana di Geotecnica 51(2)
20. Ciurleo M, Mandaglio MC, Moraci N (2021) A quantitative approach for debris flow inception and propagation analysis in the lead up to risk management. Landslides 18:2073–2093
21. Gioffrè D, Mandaglio MC, di Prisco C, Moraci N (2017) Evaluation of rapid landslide impact forces against sheltering structures. Riv Ital Geotec 51(3):64–76
22. Gullà G, Mandaglio MC, Moraci N (2006) Effect of weathering on the compressibility and shear strength of a natural clay. Can Geotech J 43(6):618–625

Community Outreach Through Soft-Type Measures in the Wake of an Initial Disaster

Atsuo Sato

Abstract Recently, natural disasters caused by linear precipitation zones in Japan have recurred in the same region shortly after the initial disaster. During recovery from disasters, many local governments are overwhelmed by the restoration of infrastructure in the public sector. However, since we do not know when the next disaster will occur, community outreach through soft-type measures needs to be considered and developed from the recovery stage based on the lessons learned from the initial disaster. This study explored the relationship between local government measures for community outreach and their agility, using the case of Takeo City, Saga Prefecture. The study also examines the local government's adaptive governance in policy-making and implementing countermeasures immediately after the initial disaster.

Keywords Disaster risk reduction · Adaptive governance · Soft-type measure · Agility

1 Introduction

In recent years, natural disasters (floods, landslides, etc.) caused by linear precipitation zones in Japan have sometimes recurred in the same area immediately after a disaster [1–4]. In many local governments, the immediate post-disaster recovery phase is dominated by the restoration of infrastructure (roads, bridges, etc.). They require enormous public financial expenditure and human resources in local governments.

However, even under such circumstances, soft measures (development of smartphones' applications for disaster prevention, etc.) for community outreach are also important to prevent and mitigate future disaster damage. Since one never knows when the next disaster will strike, community support through soft measures needs

A. Sato (✉)
Hiroshima City University, Hiroshima, Japan
e-mail: sato-a@hiroshima-cu.ac.jp

H. Hazarika et al. (eds.), *Multidisciplinary Approach for Disaster Resilience and Sustainable Technologies*, Lecture Notes in Civil Engineering 642,
https://doi.org/10.1007/978-981-96-8618-6_9

to be considered and designed from the recovery stage, based on the lessons learned from the initial disaster. Therefore, this study identified what soft measures were taken during the recovery stage of the first flood in a local government that experienced two floods in the past five years, and how these soft measures helped in the second flood. Disaster risk management might be recognized as a management cycle consisting of prevention, response, and recovery phases, rather than a one-time event. This means that the recovery phase leads to the next prevention and response phase.

In addition, the study examined governance that enabled an agile response after an initial disaster. Governance responding to situations after disasters can be referred to as adaptive governance (AG). This AG has become an important concept in dealing with uncertainty in terms of collaboration, flexibility, and learning for disaster risk reduction (DRR) [5, 7, 12]. Previous discussions on AG have focused on case studies of when AG took place. Djalante et al. classified the main characteristics of AGs into four dimensions: (1) polycentric and multilayered institutions, (2) participation and cooperation, (3) self-organization and networks, and (4) learning and innovation [7]. However, discussion on the condition where AG works well is insufficient. In this study, the condition for AG, especially AG related to (4) learning and innovation among the four dimensions above, was examined. Through the case of Takeo City, Saga Prefecture, the study explored the factors contributing to the agility in policy-making and implementation by the local government.

The Research Question (RQ) is: What was the mechanism of the city's agile response after an initial disaster (flood)? The detailed questions for the RQ are: (1) who proposed the initial idea for the measure? (2) how did the idea evolve during the policy-making and implementation process? and (3) whether were there any organizational, personnel, budgetary, or technical obstacles? The survey included information obtained during fieldwork at the Takeo Lab which was set up between Kyushu University and Takeo City in 2022 and 2023, as well as information based on an interview with the city official from the Disaster Prevention and Mitigation Division conducted in January 2023.

2 Literature Review on Adaptive Governance on Flooding

AG is recognized as an adaptation of governance for disaster risk management, climate change adaptation, and ecosystem resource management [6, 7]. Recently, many scholars have discussed AG, focusing on cases of flooding related to the four areas mentioned above. However, the AG for the prevention and mitigation of damages caused by flooding varies, and the literature shows the limitations of autonomous adaptation and maladaptation [8, 13]. Simultaneously, in some countries, there is ample room for local government leaders and officials to initiate measures as front-runners [8]. Walch identifies AG (in this case, polycentric and multilayered institutions) in the case of natural hazards (particularly cyclones) in the State of Odisha, India, and insists that traumatic shock and committed political leadership are important factors when AG is adopted in the State of Odisha [12]. However, it is not clear how leadership works for AG at the micro level.

Flooding during the COVID-19 pandemic causes people to face difficulties when evacuating and staying in evacuation centers. Some volunteers were also reluctant to help in the evacuation centers. Kumamoto Prefecture, Japan, experienced heavy rains in July 2020. Floods during the spread of the COVID-19 pandemic in Kumamoto Prefecture led to new participatory approaches and platforms among local governments, companies, and communities, which responded to disaster management with AG (in this case, polycentric and multilayered institutions) [9]. However, research has not analyzed the reason why AG was adopted, except for emphasizing the complexity and uniqueness of the situation.

3 Description of the Case of Takeo City

3.1 Overview

With global warming, floods caused by linear precipitation zones have become more frequent in the same area within a few years. Therefore, the speed and timing of soft-type measures are becoming more important. This section will analyze the soft measures that were promptly implemented by local governments in the wake of an initial disaster (flooding).

Takeo City in Saga Prefecture, located in southern Japan, experienced flood damage in 2019 and 2021, respectively. Soft-type countermeasures were considered, and preparations for implementation began during the recovery phase of the first flood disaster. Those measures were effective for the next flood in 2021. The main soft measures after the 2019 and 2021 floods will be detailed as follows. The process by which those measures were developed and implemented will also be examined.

3.2 Flooding in 2019 and Post-flooding Measures

From August 26th to 29th, 2019, heavy rainfall occurred mainly in the Northern Kyushu region, and in some places, the precipitation in August was more than twice the normal amount. In particular, a linear precipitation zone was formed and maintained over the Northern Kyushu region at dawn on the 28th, resulting to record-breaking heavy rainfall. Heavy rainfall caused rivers to overflow, flood damage, and landslides, resulting in human casualties and damage to homes and farmland, mainly in the Northern Kyushu region [10].

In Takeo City, two soft measures to reach out to the communities during the recovery phase of the 2019 floods were considered, budgeted, and promptly implemented in the process of creating a recovery plan. The first was the distribution of individual information receivers to all households, and the second was the development of a disaster prevention application for smartphones.

Even before the disaster in 2019, individual receivers were distributed to a group of people; for instance, people in charge of firefighting and disaster prevention. However, some citizens, not having individual receivers, said that they could not listen to announcements on disaster information via outside speakers during heavy rain. In the wake of the 2019 floods, Takeo City decided to distribute individual receivers to all households in Takeo City, as critical information on disaster risk reduction needs to reach everyone in Takeo City.

From 2020 to 2021, individual receivers were distributed in Takeo City, with priority given to areas affected by the 2019 floods. Some households did not want to install individual receivers, and as of January 2023, 50% of all households have installed the receivers. Swift policy-making and implementation of receiver distribution was useful for the disaster in 2021.

For the same reason as the distribution of individual receivers, a disaster prevention application for smart phones was also developed. Since young people often obtain information from their smartphones, the development of the "Takebo" disaster prevention application was initiated to target smartphone users, including young people. This disaster prevention application "Takebo" became operational in June 2021 and were useful during the torrential rainfall in August 2021.

The policy-making and implementation process of the two measures described above will be examined here in detail. That is, (1) who proposed the initial idea for the measures, (2) how the idea evolved during the policy-making and implementation process, and (3) whether there were any obstacles in terms of organization, personnel, budget, and technology.

First, the mayor presented a policy direction, that is, that important information on disaster prevention needs to reach all citizens. The staff of Takeo City's Disaster Prevention Division followed the mayor's direction and came up with ideas, which were incorporated into the Takeo City Creative Reconstruction Plan formulated in January 2020. Based on this plan, a budget was prepared during the FY2020 budgeting process, and the city proceeded to implement the policy from April 2020. During this time, the mayor encouraged and supported staff in the development and implementation of the policies. According to the staff, there were no organizational, personnel, budgetary, or technical obstacles to the development and implementation of the two policies.

3.3 Flooding in 2021 and Post-flooding Measures

A rain front remained in Kyushu for approximately a week, from August 11th, 2021, causing unusually heavy rainfall, especially in the Northern part of Kyushu, including

Saga Prefecture [11]. This caused huge flood damage to Takeo City. In the wake of the disaster in 2021, the mayor of Takeo City provided direction, that is, that places as bases for reliable evacuation should be secured, and city officials came up with the idea of establishing local evacuation centers and expanding vehicle shelters. The ideas were also listed in the Takeo City New Creative Reconstruction Plan in 2021.

When formulating the Takeo City New Creative Reconstruction Plan in 2021, there was an opportunity for officials to propose their ideas for the plan. Thus, the city hall has an atmosphere (culture) that allows officials to freely make proposals under the current city mayor's administration. In addition to the measure mentioned above, a digital inland water hazard map has been created at the direction of the mayor. A plan is also underway to establish a temporary FM station in the event of a disaster.

One different thing for public administration after the 2021 flooding from the 2019 flooding was Takeo City made the entrance hall in City Hall a one-stop service space for citizens to go through all procedures for recovery. This enabled disaster victims to go through the procedures without having to visit various departments in City Hall. This was the result of the mayor's directive to ensure that the livelihood support menu certainly reached disaster victims, and the official in charge of the project came up with an exceptional way (use of entrance hall) to realize the city mayor's direction.

3.4 Mode of Policy-Making and Implementation of Takeo City

In this section, what kind of AG is capable of implementing soft measures during the post-disaster recovery stage will be explored from the Takeo City case study. In other words, while not necessarily all local governments can immediately implement appropriate measures to overcome the initial weakness of a disaster, this study examines how Takeo City is making policy decisions and implementing them.

First, the policy-making process between the mayor and his staff will be focused on. The mayor issued directives as early as possible after the flooding occurred. The mayor did not mention the solution (specific means), but only the problem to be solved. The mayor posed to the staff in the form of a question what could be done to solve the problem. The city staff then came up with specific solutions and discussed them for implementation. The mayor supported the staff's ideas on solutions until a policy was developed and implemented. A city official from the Disaster Prevention and Mitigation Division provided a comment that this way of policy-making gives staff a sense of speed.

Second, two creative reconstruction plans, prepared in 2019 and 2021, respectively, will be focused on. Normally, people would expect a plan to be many pages long, but both are one-page long. However, it seems that what is included here will certainly be implemented. In addition, since the plans were prepared in time for budgeting, anything published in this plan was sure to be budgeted. Thus, a one-page plan was useful for the city to reduce barriers to implementation by showing both city staff and citizens what would be done after recovery. In addition, the word "creative" in the naming of the plan suggests an intent not simply to restore the area to its original state, but to make it better than it was before the disaster.

Here, as mentioned above, we learned what is important for AG from two things: (1) the policy-making process between the mayor and his staff, and (2) the Takeo City Creative Reconstruction Plans. This form of policy-making and implementation creates a sense of speed (agility), flexibility, and innovation of ideas that is very important for AG.

4 Discussion

Takeo City's case of AG is categorized as a case of learning and innovation based on the four dimensions in Sect. 1. When there are multiple ways to solve a problem, it is a good way for the mayor to ask a question and for the staff to think about how to solve the problem, which is a good way for innovation to occur. One risk of this approach is that the mayor's questions are not always appropriate. The other risk is that the staff may not always have a good answer to the mayor's question, or may not always come up with one. The immediate solution is for both the mayor and the staff to listen to the voices of more victims and experts.

5 Conclusion

When a major disaster strikes, officials tend to focus on infrastructure development, but the case of Takeo City shows that the soft measures taken immediately during the recovery phase to reach out to the community were useful for prevention and mitigation of the next disaster. The characteristics of governance that enable agile response were also examined based on the case of Takeo City. The results show that appropriate AG, as shown in Sect. 3.4, leads to agile policy-making and implementation. With climate change, public administrations are required to be agile, and there is a growing need for research on Adaptive Governance that is appropriate for such a response. Since this is only one case study, it is not possible to generalize the results, but adding more cases in the future will contribute to building on the theory.

References

1. Cabinet Office (2019) White paper disaster management in Japan. https://www.bousai.go.jp/kaigirep/hakusho/pdf/R1_hakusho_english.pdf. Last accessed 6 Jan 2023

2. Cabinet Office (2020) White Paper disaster management in Japan. https://www.bousai.go.jp/en/documentation/white_paper/pdf/R2_hakusho_english.pdf. Last accessed 6 Jan 2023
3. Cabinet Office (2021) White Paper disaster management in Japan. https://www.bousai.go.jp/en/documentation/white_paper/pdf/2021/R3_hakusho_english.pdf. Last accessed 6 Jan 2023
4. Cabinet Office (2022) White Paper disaster management in Japan. https://www.bousai.go.jp/en/documentation/white_paper/pdf/2022/R4_hakusho_english.pdf. Last accessed 6 Jan 2023
5. Choudhury M, Emdad Haque C, Doberstein B (2021) Adaptive governance and community resilience to cyclones in coastal Bangladesh: addressing the problem of fit, social learning, and institutional collaboration. Environ Sci Policy 124:580–592
6. Dietz T, Ostrom E, Stern PC (2003) The struggle to govern the commons. Science 302(5652):1907–1912
7. Djalante R, Holley C, Thomalla F (2011) Adaptive governance and managing resilience to natural hazards. Int J Disaster Risk Sci 2(4):1–14. https://doi.org/10.1007/s13753-011-0015-6
8. Fournier M, Larrue C, Alexander M, Hegger D, Bakker M, Pettersson M, Chorynski A (2016) Flood risk mitigation in Europe: how far away are we from the aspired forms of adaptive governance? Ecol Soc 21(4):49
9. Izumi T, Das S, Abe M, Shaw R (2022) Managing compound hazards: impact of COVID-19 and cases of adaptive governance during the 2020 Kumamoto flood in Japan. Int J Environ Res Public Health 19. https://doi.org/10.3390/ijerph19031188
10. Takeo city (2019) Reiwa Gannen Hachigatu Gouu Takeo-shi Saigai Kirokushi (Takeo City disaster record book on heavy rain in august 2019)
11. Takeo River Office, Ministry of Land, Infrastructure, Transport and Tourism (2021) Reiwa Sannen Hachigatsu Gouu To Sonogaiyou (Heavy rain in august 2021 and its summary). http://www.qsr.mlit.go.jp/takeo/site_files/file/kasegawa/kasesaihyouka/2021gakushikikon06.pdf. Last accessed 6 Jan 2023
12. Walch C (2019) Adaptive governance in the developing world: disaster risk reduction in the State of Odisha, India. Clim Dev 11(3):238–252. https://doi.org/10.1080/17565529.2018.1442794
13. Winter AK, Karvonen A (2022) Climate governance at the fringes: Peri-urban flooding drivers and responses. Land Use Policy 117

Gravel-Rubber Mixtures: Macro- and Micro-scale Engineering Properties

Gabriele Chiaro

Abstract Sand-rubber mixtures (SRMs) and gravel-rubber mixtures (GRMs) can be designed to have outstanding engineering characteristics. However, in comparison with the extensive investigations carried out on SRMs, comprehensive studies characterizing GRMs are very limited, making the index, physical, and mechanical properties of such synthetic materials mostly unknown. To facilitate the use of GRMs as fill materials in geotechnical applications, this paper provides a summary of the engineering properties of GRMs obtained from detailed laboratory tests (macro-scale response) and discrete element method (DEM) numerical investigations (micro-scale response) conducted at the University of Canterbury, New Zealand, over the past five years. The experimental investigations indicate that GRMs have adequate strength and compressibility properties for use as structural fills. Moreover, GRMs can be utilized also as filters for dampening vibrations and seismic waves, as the addition of rubber aggregates enhanced their energy absorption and dynamic properties. The DEM numerical study provides important insights into the fabric and force anisotropy, load-transfer mechanism, and strong-force network at the particle-to-particle scale, confirming the existence of three distinct behavioral zones for GRMs: rigid gravel-like, intermediate/dual, and soft rubber-like. Dual-behavior GRMs have high ductility that prevents abrupt failure and allows the mixtures to withstand larger plastic deformation before failure; they are suitable energy-absorbing structural fills for foundations and embankments as well as backfill materials for retaining structures.

Keywords Gravel-rubber mixtures · Direct shear · DEM · Micro-mechanics

G. Chiaro (✉)
Department of Civil and Environmental Engineering, University of Canterbury, Christchurch, New Zealand
e-mail: gabriele.chiaro@canterbury.ac.nz

H. Hazarika et al. (eds.), *Multidisciplinary Approach for Disaster Resilience and Sustainable Technologies*, Lecture Notes in Civil Engineering 642,
https://doi.org/10.1007/978-981-96-8618-6_10

1 Introduction

End-of-life tires (ELTs) are tires that have reached the end of their usable life and cannot be reused [1]. Every year, worldwide, over 1 billion ELTs reach the end of their useful life [2]; it is estimated that 50% or more of such ELTs are discarded and destined for landfills [2]. The reuse of ELT-derived recycled granulated rubber (on its own or blended with granular soils) through large-scale civil engineering projects could provide substantial environmental benefits, minimizing land usage for stockpiling, disposal costs, soil and groundwater contamination and the risk of poisonous tire fires, as well as preserving natural resources [1]. Nevertheless, less than 10% of rubber generated from ELTs is nowadays used for this purpose globally [2]. Undoubtedly, more comprehensive and well-coordinated research endeavors are necessary to expand the current database, possibly encompassing a broader range of soil-rubber blends, rubber varieties and dimensions, and novel uses.

Sand-rubber mixtures (SRMs) mechanical and physical characteristics have been the main focus of earlier research [3]. Generally, SRMs have shown satisfactory strength, reduced shear modulus, and high damping properties [4]. However, their high compressibility may result in undesirable settlement and reduced bearing capacity [5], limiting their use in a number of geotechnical applications. From a practical viewpoint, when selecting soil types and rubber sizes for SRMs, the materials cost efficiency and availability must be taken into consideration [6]. When blended with sands, to prevent the inherent segregation of mixes consisting of small and large particles [7, 8], the tire rubber should be shred into sand-size-like (i.e., ≤ 2 mm) particles. This, however, unavoidably results in high implementation costs. As a result, in the last decade, the use of coarser soil-rubber blends, including gravel-rubber mixtures (GRMs), has progressively been recommended [1, 6, 9–12].

Nevertheless, compared to SRMs, the index, physical, and mechanical properties of GRMs are still largely unknown [3]. Hence, in New Zealand, to encourage and facilitate the use of GRMs in geotechnical applications, a research activity was carried out at the University of Canterbury, over the past five years [12–18]. Specifically, three subsequent steps were undertaken by the research team led by the author: (i) a comprehensive laboratory research was carried out to characterize the engineering properties of a variety of GRMs prepared at different volumetric rubber content (VRC) and aspect ratios (AR = $D_{50,\mathrm{Rubber}}/D_{50,\mathrm{Gravel}}$) [1, 12–17]; (ii) an original three-dimensional discrete element method model (named DEM4GRM) [18] was developed, which is able to provide an accurate description of the macro-scale mechanical direct shear behavior of GRMs; and (iii) the micro-scale mechanical response of GRMs was methodically analyzed. It has been established that although most GRMs have desirable geotechnical engineering properties to be used as structural fills (e.g., effective peak friction angle $\geq 30°$), their VRC should be limited to 40% to ensure adequate bearing capacity and reduce the long-term settlement under sustained vertical loads [16]. Besides, due to possible high concentrations of percolated contaminants, the use of sand-size rubber particles in GRMs should be avoided [16, 19].

In this paper, selected laboratory test results for GRMs are presented, and the influence of adding granulated rubber on the macro-scale engineering properties (i.e., compaction, compressibility, shear strength, energy absorption, small-strain shear stiffness and shear modulus degradation properties) of GRMs is described. Then, using the newly developed DEM4GRM model [18], the force anisotropy, load-transfer mechanism, and strong-force network at the particle-to-particle scale are examined and the change in skeleton material behavior from stiff gravel-like to soft rubber-like is described.

2 Packing Density State and Skeleton Material for Granular Soft/Rigid Mixtures

Soil-rubber mixtures are peculiar binary granular mixtures made of rigid soil grains and soft rubber particles, often having different sizes and shapes. Their particle arrangement (i.e., packing density) and mechanical response are influenced by the intrinsic properties of the materials (i.e., gradation, size and shape), the percentage of large/small and soft/rigid particles in the total volume of solids, besides the size ratio of large/small particles ($D_{50,\text{Large}}/D_{50,\text{Small}}$) and the aspect ratio of soft/rigid particles ($\text{AR} = D_{50,\text{Rubber}}/D_{50,\text{Soil}}$).

According to previous studies [e.g., 4, 11, 20, 21], a binary mixture of two-size particles can have three packing states that depend on the relative proportion of small and large particles. These states are known as floating state, non-floating state, and transitional state. In the floating state, contact between large particles is prevented by small particles. In contrast, in the non-floating state, contact between large particles occurs as small particles largely fill the voids between large particles. While, in the transitional state, large particles are only partially disconnected because of small particles only partially filling the empty spaces between large particles. As a result, the densest state, which implies the minimum number of voids in the mixture, is obtained in the transitional state [22].

The packing density of a mixture can provide useful information about its mechanical behavior. However, how the load is transferred within the mixture mainly depends on the skeleton material [7, 8, 18], which is formed when particles of the same material are in contact with each other and actively transfers loads. The material that forms the skeleton (matrix material) governs the overall mechanical behavior of the mixture. For soil-rubber mixtures, two distinct matrix materials can be expected: (i) a soil matrix that behaves like stiff soil for mixtures with low VRC and (ii) a rubber matrix that exhibits a soft rubber-like response for mixtures with higher VRC. However, based on the aspect ratio (AR), dual (transition) matrixes can also be formed. Table 1 summarizes the potential soil-rubber interactions in mixtures with different VRC and AR [3], along with distinct skeleton materials, particle level phenomena, and expected macro-scale responses.

Table 1 Schematic summary of soil-rubber interaction in mixtures with different VRC and AR (modified from Kim and Santamarina [7] and Tasalloti et al. [15])

VRC (%)	Skeleton material			Remarks
	AR $\ll$ 1	AR $\approx$ 1	AR $\gg$ 1	
0	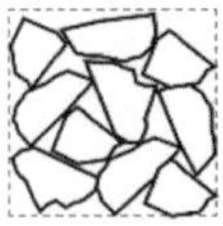	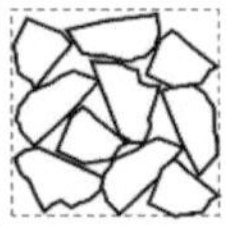		Rigid soil skeleton
20	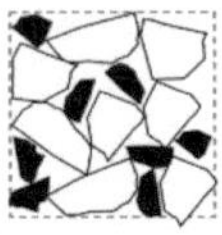		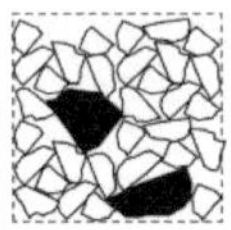	Soil-controlled stiffness; rubber may prevent buckling of soil columns; **AR $\ll$ 1**: segregation if rubber can pass through soil pores
40	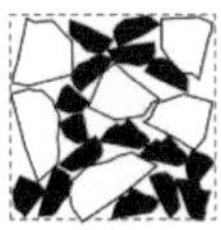	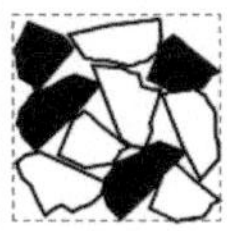	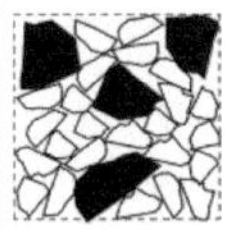	**AR $\ll$ 1**: Transaction mixture. Rubber separates soil contacts at low pressures; soil contacts may form at large pressures **AR $\gg$ 1** and **AR $\approx$ 1**: Rubber forms percolating skeleton. There is soil-soil grain interaction
60				**AR $\ll$ 1** and **AR $\approx$ 1**: Soil forms percolating skeleton. There is rubber-rubber particle interaction; **AR $\gg$ 1**: Soil separates rubber contacts
80	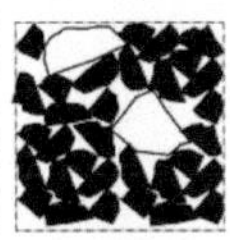			Rubber-controlled stiffness; **AR $\gg$ 1**: Segregation if soil can pass through rubber pores
100				Soft rubber skeleton

In the case of AR $\ll$ 1 (i.e., $D_{50,\mathrm{Rubber}} \ll D_{50,\mathrm{Soil}}$), rubber particles have a tendency to fill the empty spaces between soil grains in mixtures that have low VRC. However, as the VRC increases, the rubber particles separate and replace the soil grains [7, 8]. Mixtures with intermediate VRC values may develop soil-to-soil contacts when the confining pressure increases. As a result, the transition mixtures behave like rubber at low confining pressure and like soil at high confining pressure.

When AR $\gg$ 1 (i.e., $D_{50,\mathrm{Rubber}} \gg D_{50,\mathrm{Soil}}$), rubber particles have a tendency to float within the soil skeleton in mixtures with low VRC. For intermediate VRC values, the behavior of the mixtures remains consistent at all stress levels and there is no transition from rubber-like to soil-like behavior when the confining pressure increases [7, 8].

In the case of AR ≈ 1 (i.e., $D_{50,\text{Rubber}} \approx D_{50,\text{Soil}}$), for VRC $\leq 60\%$, mixtures behave like those with AR $\gg 1$; in this case, rubber forms peculated structures and soil-soil grain interaction exists. However, for VRC $> 60\%$, mixtures behave like those with AR $\ll 1$; here, the soil forms percolated structures and rubber particles interact with each other.

3 Laboratory Study on the Macro-scale Engineering Properties of GRMs

3.1 Materials and Test Program

In this paper, GRMs with aspect ratio AR $\approx$ 0.3, 0.6, and 1.0 are examined. They were produced by mixing rounded pea gravel with three different sizes of granulated recycled tire rubber (Fig. 1). The index properties of the gravel and recycled rubber particles are summarized in Table 2.

Several mixtures were prepared at various volumetric rubber content (VRC), which is defined here as follows:

Fig. 1 Gravel and rubber tested in the laboratory

Table 2 Properties of gravel and rubber used in this study

Material	Gravel	Recycled tire rubber		
Symbol	G	RL	RM	RS
Mean diameter D_{50} (mm)	5.67	5.35#	3.23	1.61
Specific gravity G_s	2.71	1.15	1.15	1.14
Maximum dry unit weight γ_{dry} (kN/m^3)+	17.2	6.37	6.22	5.91

+By Proctor impact compaction test ($E = 600$ kJ/m^3)
#Previously reported as $D_{50} = 3.7$ mm in Tasalloti et al. [13, 15] and Chiaro et al. [14]

$$\text{VRC} = \frac{V_{\text{Rubber}}}{V_{\text{Gravel}} + V_{\text{Rubber}}} \times 100 \ (\%) \tag{1}$$

where V_{Rubber} and V_{Gravel} are the volume of the rubber particles and the volume of gravel grains in the mixture, respectively.

All laboratory tests were conducted on dry specimens that were prepared at a degree of compaction of 90–95% using the tamping method, as per the standard Proctor impact compaction tests mentioned in Sect. 3.2. To prevent segregation in the specimens, any vibration was minimized and granular flow was avoided.

As described hereafter, in the laboratory the macro-scale engineering properties of GRMs were evaluated by conducting one-dimensional compression tests, direct shear tests, bender element tests, and small-strain cyclic triaxial tests.

3.2 *Test Results*

Specific gravity. Typical experimental results showing the variation of G_s with VRC are reported in Fig. 2. It can be seen that, irrespective of the AR, the G_s of GRMs tend to decrease linearly with increasing VRC. Therefore, if the G_s of both the parent materials are known, the linear expression shown in Eq. (2) can be used to estimate G_s of the mixtures:

$$G_{s,\text{Mix}} = (G_{s,\text{Gravel}} - G_{s,\text{Rubber}})\left(1 - \frac{\text{VRC}}{100}\right) + G_{s,\text{Rubber}} \tag{2}$$

Compaction characteristics and packing density state. To ensure satisfactory performance of structures made of or placed on GRMs, it is necessary to properly

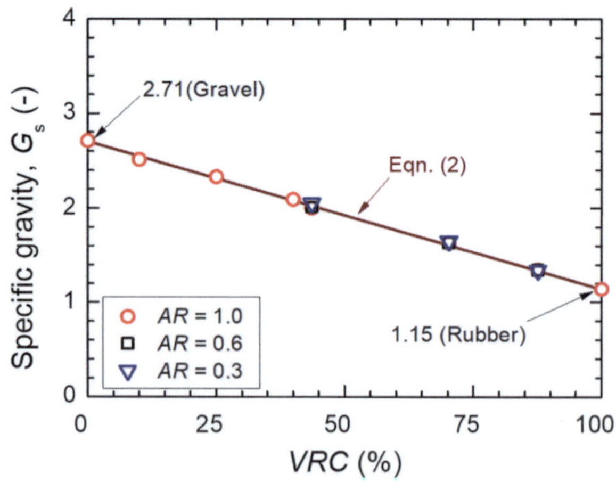

Fig. 2 Variation of specific gravity with VRC for GRMs with AR = 1.0, 0.6, and 0.3 (experimental data points from [1, 16])

evaluate the compaction characteristics and understand the packing density state of GRMs.

Experimental evidence suggests that the Proctor compaction test produces a higher maximum dry density/unit weight for SRMs [23, 24] rather than the vibratory table method, which is usually more suitable for hard-grained cohesionless soils (i.e., sands and gravels) [25]. This is mainly due to the ability of rubber particles to absorb most of the energy produced by vibrations. In this study, it was confirmed that the Proctor compaction test is also appropriate for GRMs [15]; moreover, it was established that due to the high permeability of free-draining GRMs, the moisture content does not influence the evaluation of the maximum dry density/unit weight of GRMs in the Proctor compaction test [15].

The variation of maximum dry unit weight (γ_{d}) with VRC is presented in Fig. 3a–c. It can be seen that the γ_{d} of the mixtures decreases by increasing VRC; this is essentially because the rubber particles are much lighter than the gravel ones (Fig. 2). Interestingly, for GRMs with AR = 1.0, the experimental data points follow a linear trend (Fig. 3a) that can be defined by Eq. (3):

$$\gamma_{d,\mathrm{Mix}} = \left(\gamma_{d,\mathrm{Gravel}} - \gamma_{d,\mathrm{Rubber}}\right)\left(1 - \frac{\mathrm{VRC}}{100}\right) + \gamma_{d,\mathrm{Rubber}}. \tag{3}$$

Nevertheless, as AR becomes smaller than 1.0, the data points start to deviate from the linear trend (Fig. 3b, c), as the packing density state (i.e., particle arrangement) significantly changes. That is, at AR = 1.0, the size of the rubber particles and that of the gravel grains is the same, and thus the particles of rubber cannot occupy the empty spaces between the gravel grains, even at lower VRC (Fig. 3a); hence, the addition of rubber to gravel does not alter the void ratio (Fig. 3d) or packing density state. In contrast, at AR = 0.3 (for VRC $\leq 40\%$), small rubber particles can easily fill in the voids between larger gravel grains (Fig. 3b), resulting in a smaller void ratio (Fig. 3d) and, thus, in an increase in the γ_{d} of the mixtures.

Compressibility. Design considerations require taking into account material compressibility, which refers to the ability of a material to increase or decrease in volume when it is subjected to an applied load. In the case of conventional granular soils, which have rigid particles like sand and gravel, any volume change occurs due to the rotation, movement, and rearrangement of non-compressible grains [7]. Nevertheless, the compressibility of GRMs is significantly different from that of granular soils because of the elastic modulus differences between the gravel grains (rigid) and the rubber particles (soft). Also, individual rubber particles do not undergo much volume change under normal stress conditions (this is due to Poisson's ratio of pure rubber being $\nu \approx 0.5$), and distortion is the primary mechanism that occurs in pure rubber particle assembles.

To evaluate the compressibility of GRMs, one-dimensional (1-D) compressibility tests were carried out using a medium-size compression cell (specimen dimensions: inner diameter = 250 mm; height = 150 mm). During the tests, 2-h creep was applied at different incremental loading stages. This duration was sufficient to achieve

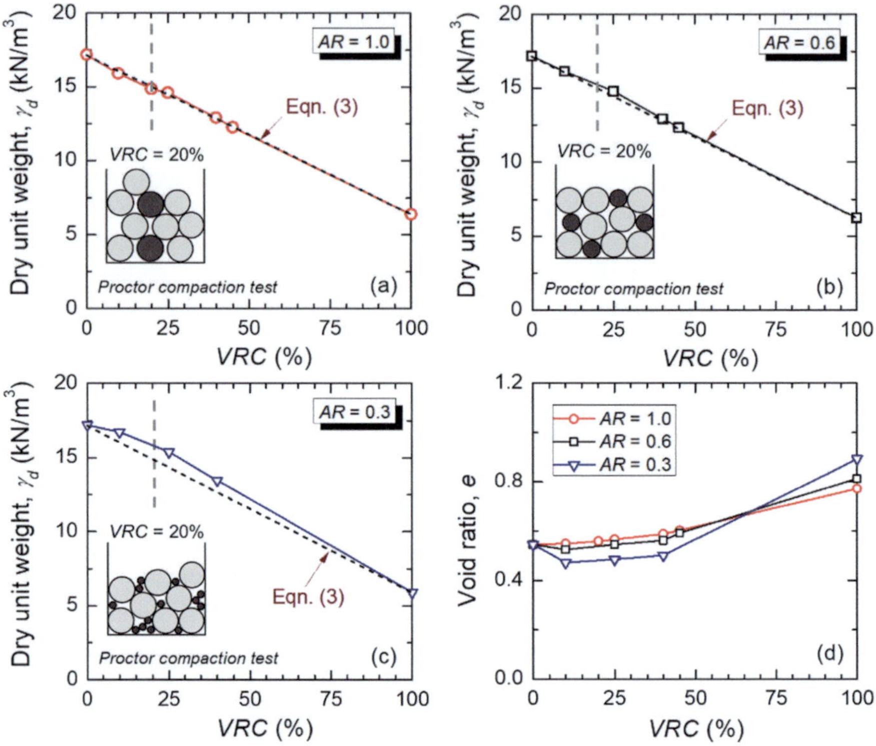

Fig. 3 Typical results of Proctor compaction tests on GRMs: **a–c** variation of maximum dry unit weight with VRC; and **d** variation of minimum void ratio with VRC

negligible settlement under sustained vertical stress. The measured vertical stress was corrected for the loss of the transferred load (experimentally determined), which was mainly due to soil-wall friction.

Figure 4a shows the stress–strain relationship of GRMs under zero lateral strain loading, also known as k_0 conditions. It can be observed that an increase in 1-D compression occurs with increasing VRC, for any specified vertical shear stress level. Also, it is evident that, under any specific vertical stress level, GRMs with 10% VRC performed better than those with 25 and 40% VRC as they developed considerably smaller vertical strains. Moreover, GRMs consisting of large rubber particles (AR = 1.0) were considerably less compressible than those made of small rubber particles (AR = 0.3).

Figure 4b shows the variation of the secant drained constrained modulus ($M = \sigma_v/\varepsilon_v$) with VRC computed at 50, 100, and 200 kPa vertical stress. It can be seen that M drastically decreases with increasing VRC and much less with increasing AR. The data suggest that as the VRC increases, the mixture stiffness gradually reduces and the compressibility increases; this is a clear indication that at lower VRC the mixtures support the load mainly through the gravel skeleton while at larger VRC

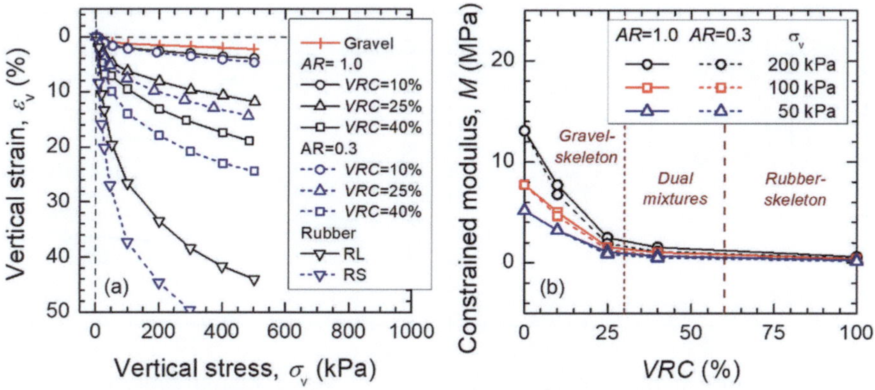

Fig. 4 Results of 1-D compression tests with 2-h creep for GRMs with AR = 1.0 and 0.3: **a** vertical stress–vertical strain relationships; and **b** variation of constrained modulus with VRC

the load is transferred primarily through the rubber skeleton. This aspect is discussed in more detail in Sect. 4.

Shear strength. The estimate of the shear strength was obtained using a medium-size direct shear box (specimen dimensions: 100 mm × 100 mm in cross-section and 53 mm in height). The specimens were subjected to normal stress levels of 6.5, 30, 60, and 100 kPa, while the horizontal displacement rate was maintained at 1 mm/min. The setup for the test is described in detail in [12–17].

The data in Fig. 5 present the typical shear strength and volumetric behaviors of GRMs with AR = 0.6. In Fig. 5a, it is observed that when rubber is added to the mixture, the stiffness and peak stress of the mixtures reduce under the same normal stress applied (i.e., σ_n = 100 kPa). This is because rubber materials have a lower shear modulus and shear strength in comparison with gravel. Furthermore, as VRC increases, the post-peak strain softening is reduced, and the horizontal displacement corresponding to the peak stress is larger. This indicates that the response of GRMs changes from brittle to ductile, which is advantageous for utilizing rubber in structural fills for foundations and embankments, as well as in backfill materials for retaining structures. The higher ductility can prevent abrupt failure and allows GRMs to undergo larger plastic deformation before failure. As expected, as VRC increases, the GRMs become more contractive/less dilative (Fig. 5b).

Similar responses were observed for all mixtures (AR = 0.3, 0.6, and 1.0) irrespective of the normal stress level applied (as reported in [12–17]). The typical variation of peak shear stress (τ_{max})—evaluated at failure for brittle materials and at large horizontal displacement (i.e., 15 mm) for ductile materials—with normal stress (σ_n) is shown in Fig. 6a for GRMs with AR = 0.6. The data indicate that GRMs are essentially cohesionless materials. It is also evident that τ_{max} increases with increasing σ_n and decreases with decreasing VRC. Instead, as shown in Fig. 6b, the effects of AR on the GRM strength appear to be insignificant.

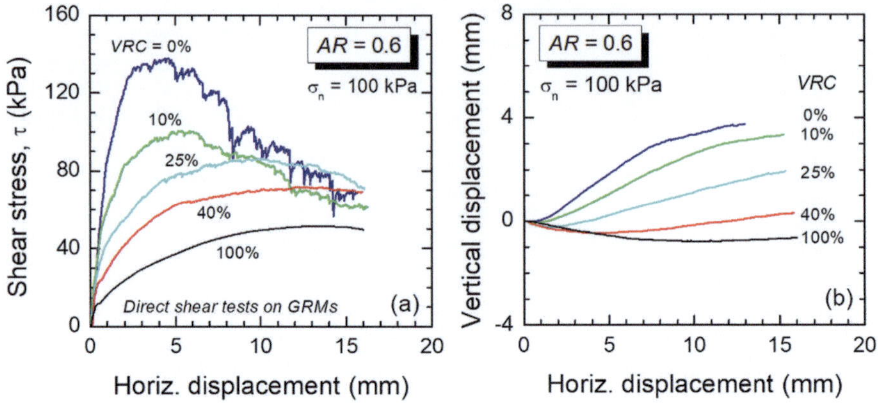

Fig. 5 Typical direct shear results for GRMs with AR = 0.6 sheared at 100 kPa normal stress: **a** strength behavior and **b** volumetric response

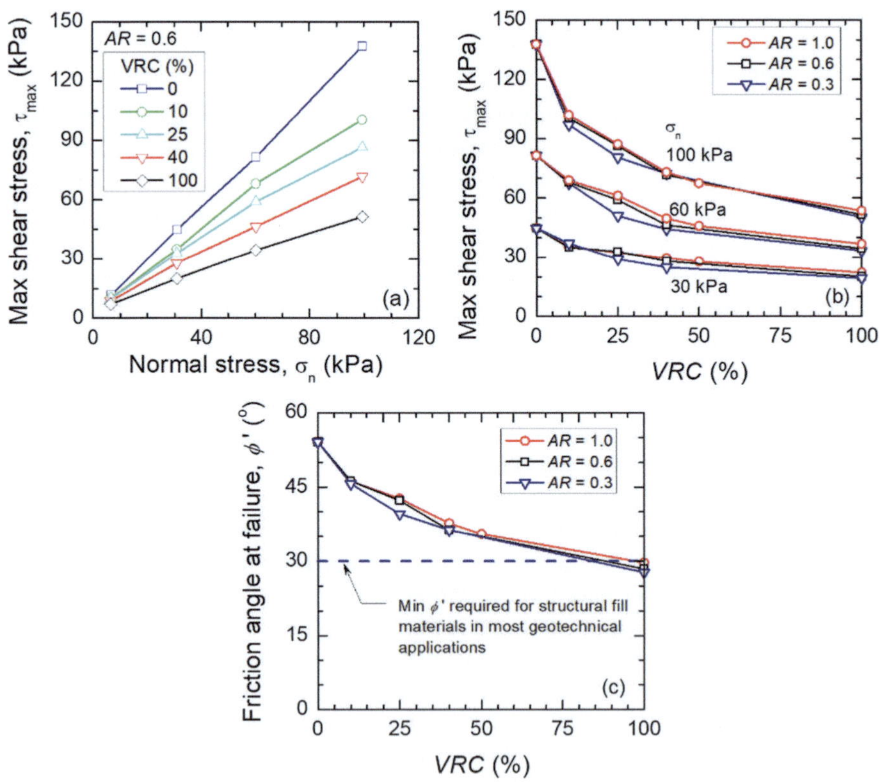

Fig. 6 Summary of the direct shear test results on GRMs: **a** typical maximum shear stress versus normal stress relationships; **b** variation of maximum shear stress; and **c** Mohr–Coulomb friction angle

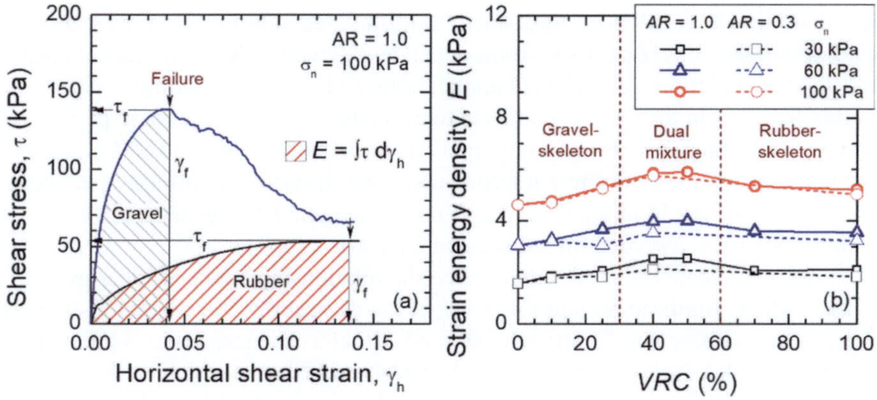

Fig. 7 Strain energy density for GRMs: **a** definition and **b** variation of SED with VRC and normal stress in direct shear tests

In Fig. 6c, the Mohr–Coulomb friction angle (ϕ') values are reported for all mixtures. It can be observed that ϕ' decreases significantly with increasing VRC from about 54° for the gravel to 29° for the rubber; whereas, the effect of AR is negligible. Remarkably, most of the GRMs have adequate strength (i.e., $\phi' \geq 30°$) making them suitable structural fill materials for typical geotechnical applications [26].

The trends for ϕ' shown in Fig. 6c are consistent with those shown in Fig. 4b for *M* confirming a transition from gravel skeleton to rubber skeleton as VRC increases. This is further discussed in Sect. 4.

Energy absorption. In this study, the concept of strain energy density (*E*) [27] is used to investigate the energy absorption characteristics of GRMs. As depicted in Fig. 7a, *E* is the integral of the area beneath the shear stress–strain curve until failure (i.e., the state where the specimen reaches the peak stress, τ_f).

According to the summary plot displayed in Fig. 7b, an increase in normal stress leads to an increase in *E*, indicating that higher normal stress levels result in more work input being absorbed by the specimens. The addition of rubber to gravel results in a noticeable increase in *E*, reflecting the high energy absorbed by rubber aggregates. However, it appears that the peak *E* value is achieved at VRC = 40–50% for any specified normal stress level applied, and no further increase in *E* is observed beyond this point. This is due to the decrease in peak strength compromising the increase in ductility, which results in failure occurring at a larger shear strain level (as depicted in Fig. 5).

As the addition of rubber aggregates to gravel improves the energy absorption and dynamic properties, GRMs have been proposed for use as filters for dampening vibrations and seismic energy waves in geotechnical seismic isolation systems [9, 28–30].

Small-strain shear stiffness and shear modulus degradation. The estimation of small-strain stiffness (G_{max}) and strain-dependent dynamic characteristics of GRMs is crucial in designing geotechnical applications under seismic loading conditions.

Bender element tests were conducted to evaluate G_{max}. The tests were performed in a triaxial apparatus at confining pressures (σ_c') of 20, 40, 60, 80, 100, and 200 kPa. Shear waves of different frequencies were transmitted through specimens, which were 60 mm in diameter and 130 mm in height. The signals received by an oscilloscope were recorded for subsequent analysis.

In order to create the curves that define the relationship between shear modulus (G) and shear strain amplitude (γ), a set of small-strain cyclic triaxial tests were carried out at different σ_c' of 40, 60, 100, and 200 kPa. The tests were conducted on specimens with a diameter of 70 mm and a height of 170 mm, which were subjected to stress-controlled cycles of shear loading at a frequency of 0.05 Hz. The deviatoric stress amplitude was varied at each σ_c' level to determine G at different shear strain (γ) levels.

The results of bender element tests reported in Fig. 8a indicate that G_{max} of GRMs decreases as VRC increases, while it increases as σ_c' increases. This is in agreement with previous studies on both GRMs and SRMs [31]. The relationship between G_{max} and σ_c' can be mathematically expressed as a power law relationship (Eq. 4):

$$G_{max} = G_{100}\left(\frac{\sigma_c'}{100}\right)^{0.5} \tag{4}$$

where G_{100} is the small-strain shear stiffness at the reference $\sigma_c' = 100$ kPa.

The graph in Fig. 8b displays the degradation curves of the shear modulus (G) at $\sigma_c' = 100$ kPa. It is evident that the inclusion of rubber in gravel hinders the degradation of the shear modulus during cyclic loading. This finding is important as it suggests that GRMs perform better than pure gravel under cyclic loading, as they maintain their initial stiffness and dynamic properties over a wider range of shear strains. This is because the more deformable rubber particles have the ability to rebound and dissipate energy.

Irrespective of the level of σ_c' and VRC, all the G-γ curves follow a similar pattern that can be mathematically described by a hyperbolic model Eq. (5):

$$\frac{G}{G_{max}} = \frac{1}{1 + (\gamma/\gamma_{0.5})^{a}} \tag{5}$$

where $\gamma_{0.5}$ is the reference shear strain evaluated at $G/G_{max} = 0.5$, and α is a material fitting coefficient. For the GRMs examined in this study, the $\gamma_{0.5}$ and α values can be obtained by the following linear expressions: $\gamma_{0.5} = 0.0004$ VRC + 0.0121 and $\alpha = 0.0044$ VRC + 0.5706 [17]. As shown in Fig. 9, using Eq. (5), a set of G/G_{max} curves that are valid for confining pressure up to 200 kPa can be obtained.

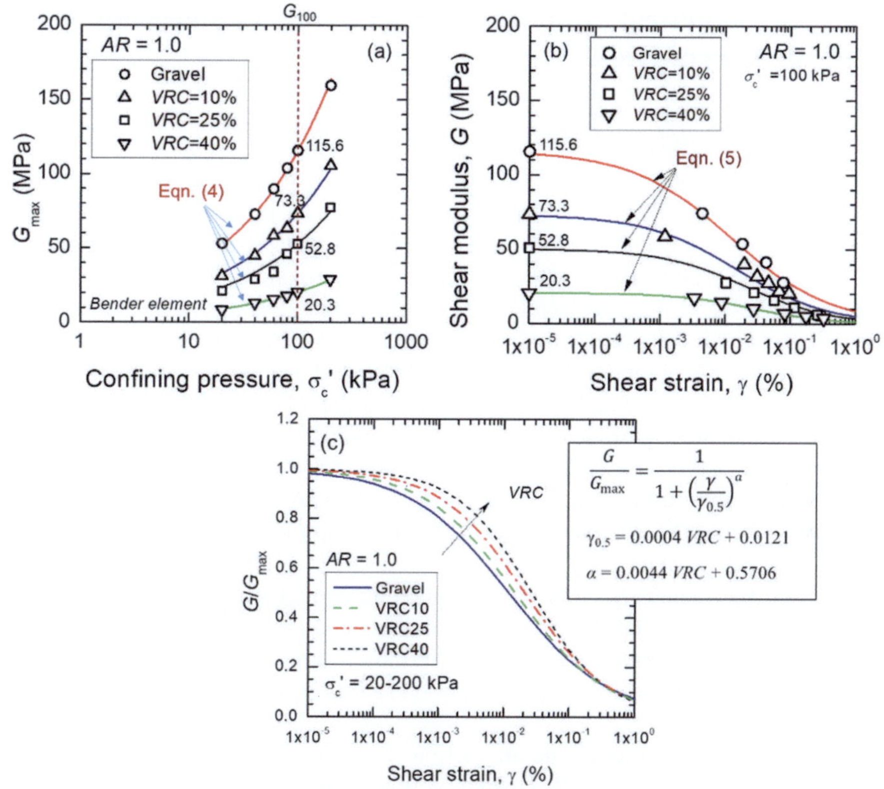

Fig. 8 Strain-dependent properties of GRMs: **a** small-strain shear stiffness evaluated by bender element tests; **b** shear modulus degradation evaluated by cyclic triaxial tests; and **c** G/G_{max} versus shear strain curves for GRMs derived using Eq. (5)

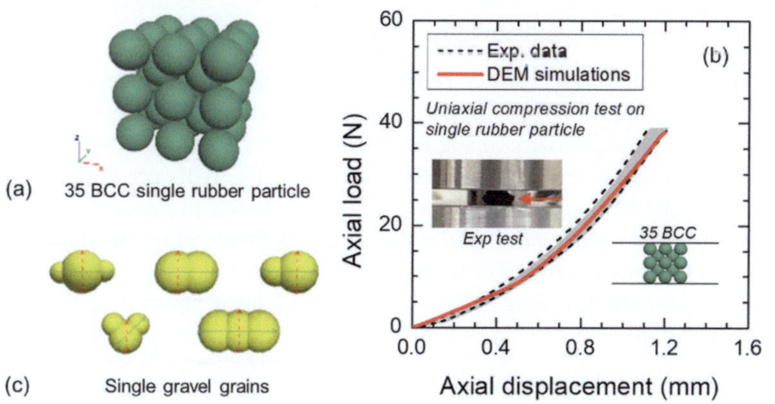

Fig. 9 DEM4GRM model characteristics: **a** 35 BCC ball single rubber particle; **b** 35 BCC single rubber particle response in uniaxial compression test; and **c** single gravel grains

4 DEM Numerical Study on the Micro-scale Engineering Properties of GRMs

4.1 DEM Model for GRMs

The discrete element method (DEM) is a popular tool used to study how soil particles interact with each other. It is helpful in developing an in-depth understanding of both the macroscopic and microscopic mechanical performance of granular materials [32]. Because GRMs have a discrete and granular nature, DEM is an ideal numerical modeling method to scrutinize their micro-mechanical behavior [18].

A 3D hybrid DEM model for GRMs was developed using the Particle Flow Code 3D program (PFC3D) [33]. This model is called the DEM4GRM model [18] and takes advantage of the particle rigidity approach applicable to hard-grained soils [34] and the ability of reproducing the elastic rubber response using body-centered-cubic (BCC) deformable particles [35, 36]. It allows for the examination of the micro-scale mechanical direct shear behavior of GRMs under varying VRC and normal stress levels.

The PFC3D framework employs an explicit numerical scheme to track the interaction of single particles individually and their contacts with adjacent particles [33]. Additionally, contact forces are determined based on the particle rigidity approach, which evaluates the overlap between contacting bodies without any particle deformation [32].

In the DEM4GRM model, the interactions at each contact are defined using the Hertz-Mindlin contact law [37]. This law follows a nonlinear contact force-overlap relationship, which is controlled by a user-defined initial G and a dashpot component for damping. Table 3 provides the DEM input parameters that define the contacts. Since the gravel grains and rubber particles have distinct engineering properties, the surface property inheritance functionality in PFC3D was used to define the hybrid contact properties between them [33].

DEM modeling of single rubber particle. A model was developed to accurately account for the compressive volumetric behavior of rubber. The model consists of a single rubber particle that is made up of a cluster of 35 balls, which are packed together in a body-centered cubic (BCC) structure. The balls are bonded together by linear parallel bonds [38], as shown in Fig. 9a. When in contact with external particles, the inter-cluster non-bond contacts are defined using the Hertz-Mindlin model [37]. As shown in Fig. 9b, the results of a series of uniaxial compression tests on single rubber particles were used to determine the elastic parameters of the model. For more information on the single rubber particle model, readers can refer to Chew et al. [18].

DEM modeling of single gravel particle. To account for grain size and shape variability, the single gravel grains were modeled using a combination of simple clump shapes similar to those utilized by Garzia and Bray [39] (Fig. 9c).

Table 3 Input parameters for the DEM4GRM model [18]

Item	Parameter	Value
Gravel	Particle density (kg/m^3)	2710
	Coefficient of friction	0.72
	Shear modulus (MPa)	90
	Poisson's ratio	0.3
Wall	Coefficient of friction	0.70
	Shear modulus (GPa)	80
	Poisson's ratio	0.25
Global	Damping coefficient	0.2
Rubber bonds	Shear modulus bonded spheres (MPa)	12
	Bond elastic modulus (MPa)	20
	Normal-to-shear stiffness ratio	4.0
	Tensile strength (kPa)	30E200
	Shear strength (kPa)	30E200
	Critical damping ratio	0.25
Rubber	Particle density (kg/m^3)	1140
	Coefficient of friction	0.27

DEM modeling of GRM assembly. The specimens of GRMs were created using the overlapping method [36, 40] and the material particle size distribution curves from laboratory experiments (Fig. 1). The specimens were compressed until the desired compaction degree was achieved. The normal stress was applied through a force exerted on the top plate. Subsequently, the specimens were sheared by moving horizontally the lower box at a rate of 0.002 m/s, which ensured a quasi-static condition. To minimize k_0 effects, frictionless lateral walls were used. Figure 10a displays typical GRM specimens that were analyzed using the DEM4GRM model.

The reference material's model parameters were determined through simulations of gravel (0% VRC) and rubber (100% VRC) experimental direct shear behavior. The simulations were used to fit the shear and volumetric responses that best matched the laboratory results. No supplementary model parameters were necessary for the other mixtures. In fact, the shear and volumetric behaviors in the simulations were governed by the VRC of the mixtures and the specified normal stress applied to the specimen.

4.2 Simulation Results

Macro-mechanical response. In Fig. 10b, c, typical DEM simulation results are presented for GRMs sheared at $\sigma_n = 100$ kPa. It is evident that there is a high level of

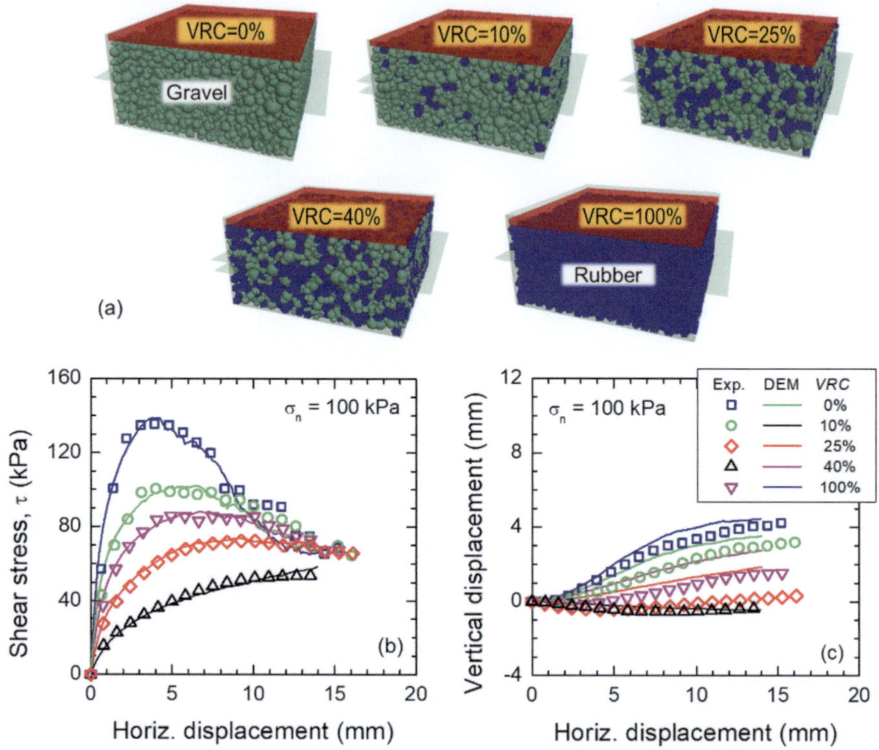

Fig. 10 DEM4GRM model outputs: **a** typical GRM specimens; **b** simulated versus exp. observed shear strength response; and **c** simulated versus exp. observed volumetric behavior

consistency between the experimental test results and the model simulations, regardless of the VRC. The stress and volumetric behaviors of the mixtures were captured effectively by the model, even though no specific model parameters were defined for the mixtures. The model was able to accurately represent key characteristics of both brittle/dilative (rigid gravel-like) and ductile/compressive (soft rubber-like) mixture responses with varying VRC. As described in Chew et al. [18], similar simulation results were obtained for the case of $\sigma_n = 30$ and 60 kPa.

Micro-scale fabric and anisotropy. To describe the fabric and force anisotropy, GRM-specific equations were developed [18] and used in conjunction with the methodology proposed by Rothenburg and Bathurst [41].

Figure 11 shows the variation of the force anisotropy coefficients (a, a_n, a_t) during the shearing process evaluated for GRMs (VRC = 25% at $\sigma_n = 60$ kPa). It is important to mention that, although not presented here, very similar trends were obtained at 30 and 100 kPa normal stress [18]. Due to page limitations, the polar histograms of the fabric, normal force, and tangential force distributions are not reported in this paper but can be found in Chew et al. [18].

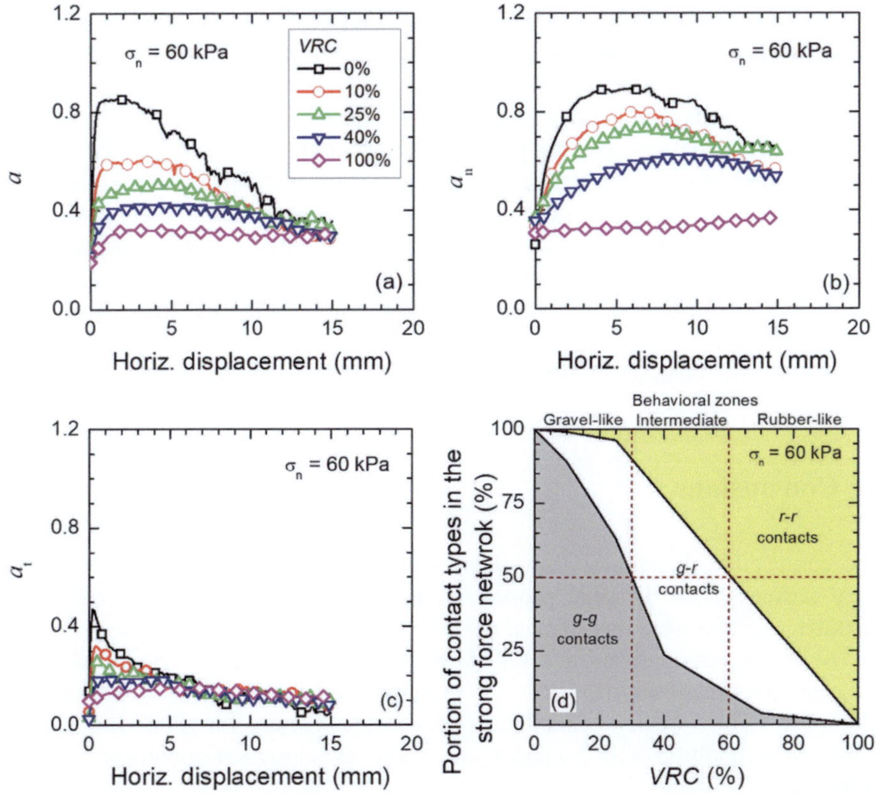

Fig. 11 Variation of force anisotropy coefficients **a** contact force distribution; **b** normal force distribution; **c** tangential force distribution; **d** strong-force network characteristics and identified behavioral zones for GRMs in direct shear tests

It is observed that the micro-scale fabric characteristics of the rubber specimens (VRC = 100%) are significantly different from those of the gravel specimens (VRC = 0%). The fabric of the mixtures appears to be similar to either gravel or rubber based on the VRC. Specifically, when VRC is equal to 40%, the trends of a, a_n, a_t begin to resemble those of rubber, which indicates that the mixtures with VRC $\geq$ 40% may exhibit a soft-like rubber-dominated behavior. Conversely, the peaks in the trends of a, a_n, a_t are noticeable up to VRC = 25%, suggesting that the mixtures with VRC $\leq$ 25% may exhibit a rigid-like gravel-dominated behavior. This observation is confirmed by analyzing the strong-force network, which is described below.

Force network and behavioral zones. It is well-known that the contact force network is divided into the strong-force network and the weak-force network [42]. The strong-force network comprises contact forces greater than the average force in the assembly. It represents the load-bearing force chains that form during shearing and is responsible for the micro-scale fabric and force anisotropy [43–45].

In Fig. 11d, the distribution of gravel-gravel (*g-g*), gravel-rubber (*g-r*), and rubber-rubber (*r-r*) contacts in the strong-force network is shown for different GRMs. The data shows that up to VRC = 30%, *g-g* contacts are responsible for more than half of the strong contacts, and the percentage of *r-r* contacts is negligible. This leads to a brittle/dilative behavior typical of gravel skeleton GRMs. However, for VRC > 30%, the percentage of strong *r-r* contacts increases significantly while the percentage of strong *g-g* contacts drops rapidly. At VRC = 40%, both *r-r* and *g-g* contacts carry around 25% of the load each, whereas the remaining 50% load is supported by *g-r* contacts. This produces an intermediate (dual) behavior where the material displays primarily a brittle/dilative behavior during shearing. When VRC exceeds 60%, *r-r* contacts sustain most of the strong-force chains, and the behavior of the GRMs becomes ductile/contractive, which is typical of rubber skeleton GRMs.

5 Conclusions

The compaction, compressibility, shear strength, energy adsorption, small-strain shear stiffness, and dynamic properties engineering properties of gravel-rubber mixtures (GRMs) were investigated through comprehensive laboratory tests.

The test results indicate that GRMs have adequate strength (friction angle $\geq$ 30°) and compressibility (if VRC $\leq$ 40%) properties to be used as structural fills. Moreover, GRMs are also suitable as filters for dampening vibrations and seismic waves, as the addition of rubber aggregates improves energy absorption and dynamic properties.

Irrespective of the normal stress level and rubber particle size (or aspect ratio, AR), the response of GRMs changes from brittle to ductile with the addition of rubber in the mixtures. This is advantageous in structural fills for foundations and embankments as well as in backfill materials for retaining structures. The higher ductility can prevent abrupt failure and allow GRMs to undergo larger plastic deformation before failure.

The DEM numerical study conducted using the newly developed DEM4GRM model made it possible to examine the fabric and force anisotropy, load-transfer mechanism, and strong-force network at the particle-to-particle scale of GRMs during direct shear loading conditions. Based on the strong-force network analysis, these distinct behavioral zones were identified for GRMs:

- gravel-like rigid behavior zone: VRC $\leq$ 30%;
- intermediate/dual-behavior zone: 30% < VRC < 60%; and
- rubber-like soft behavior zone: VRC $\geq$ 60%.

Such behavioral zones are useful to explain the change in the mechanical behavior from brittle to ductile observed in the laboratory test and associated with the skeleton material of the mixtures.

Acknowledgements The Author would like to acknowledge the research support provided by the Ministry of Business, Innovation and Employment of New Zealand (MBIE Smart Ideas Research

Grant No. 56289). A significant portion of the content has already been discussed in past issues of the NZ Geomechanics News, Soils and Foundations, Construction and Building Materials, Sustainability, Applied Sciences, Lecture Notes in Civil Engineering, and Springer Nature; salient contents from these previous publications are reproduced here with kind permission from the respective sources. The Author is thankful for contributions from Ali Tasalloti, Laura Banasiak, Alessandro Palermo, Gabriele Granello, Arjun Murali, Kevin Chew, Jayan S. Vinod, Khrisna Allulakshmi, and numerous UG students. The Author is also grateful to Sean Rees and Siale Faitotonu for their assistance during the laboratory work. The IT logistics support provided by Olive Dalton and Paul Strange is also greatly appreciated.

References

1. Chiaro G, Tasalloti A, Banasiak L, Palermo A, Granello G, Rees S (2020) Sustainable recycling of end-of-life tyres in civil (geotechnical) engineering applications: turning issues into opportunities in the New Zealand context. NZ Geomech News 99:38–47
2. WBCSD (2008) Managing end-of-life tires. WBCSD, Geneva, Switzerland
3. Tasalloti A, Chiaro G, Murali A, Banasiak L (2021) Physical and mechanical properties of granulated rubber mixed with granular soils—a literature review. Sustainability 13(8):4309
4. Mashiri MS, Vinod JS, Sheich MN, Tsang HH (2015) Shear strength and dilatancy behaviour of sand–tyre chip mixtures. Soils Found 55(3):517–528
5. Dhanya JS, Boominathan A, Banerjee S (2020) Response of low-rise building with geotechnical seismic isolation system. Soil Dyn Earthq Eng 136:106187
6. Hazarika H, Abdullah A (2016) Improvement effects of two and three dimensional geosynthetics used in liquefaction countermeasures. Jpn Geotech Soc Spec Publ 2(68):2336–2341
7. Kim HK, Santamarina JC (2008) Sand-rubber mixtures (large rubber chips). Can Geotech J 45:1457–1466
8. Lee JS, Doods J, Santamarina JC (2007) Behavior of rigid-soft particle mixtures. J Mater Civ Eng 19(2):179–184
9. Pitilakis D, Anastasiadis A, Vratsikidis A, Kapouniaris A, Massimino MR, Abate G, Corsico S (2021) Large-scale field testing of geotechnical seismic isolation of structures using gravel-rubber mixtures. Earthq Eng Struc Dyn. https://doi.org/10.5281/zenodo.4557538
10. Hazarika H, Pasha SMK, Ishibashi I, Yoshimoto N, Kinoshita T, Endo S, Karmokar AK, Hitosugi T (2020) Tire-chip reinforced foundation as liquefaction countermeasure for residential buildings. Soils Found 60(2):315–326
11. Pasha SMK, Hazarika K, Yoshimoto N (2019) Physical and mechanical properties of gravel-tire chips mixture (GTCM). Geosynth Int 26(1):92–110
12. Chiaro G, Palermo A, Banasiak LJ, Granello G (2019) Direct shear behaviour of gravel-granulated tyre rubber mixtures. In: Proceedings of the 13th ANZ geomechanics conference, Perth, Australia, pp 221−226
13. Tasalloti A, Chiaro G, Palermo A, Banasiak LJ (2020) Effect of rubber crumbs volumetric content on the shear strength of gravelly soil in direct shear apparatus. Geotech Spec Publ GSP 319:259–266
14. Chiaro G, Palermo A, Granello G, Tasalloti A, Banasiak LJ (2021) Reuse of waste tires to develop eco-rubber seismic-isolation foundation systems: preliminary results. Lect Notes Civ Eng 144:159–169
15. Tasalloti A, Chiaro G, Banasiak L, Palermo A (2021) Experimental investigation of the mechanical behaviour of gravel-granulated tyre rubber mixtures. Constr Build Mater 273:121749
16. Tasalloti A, Chiaro G, Murali A, Banasiak L, Palermo A, Granello G (2021) Recycling of end-of-life tires (ELTs) for sustainable geotechnical applications: a New Zealand perspective. Appl Sci 11(17):7824

17. Chiaro G, Tasalloti A, Palermo A, Banasiak L (2023) Small-strain shear stiffness and strain-dependent dynamic properties of gravel-rubber mixtures. In: Proceedings of the 17th symposium earthquake engineering, India, Lecture Notes in Civil Engineering, vol 331. pp 467−477
18. Chew K, Chiaro G, Vinod JS, Tasalloti A, Allulakshmi K (2022) Direct shear behavior of gravel-rubber mixtures: discrete element modeling and microscopic investigations. Soils Found 62(3):101156
19. Banasiak LJ, Chiaro G, Palermo A, Granello G (2021) Environmental implications of the recycling of end-of-life tires in seismic-isolation foundation systems. Lect Notes Civ Eng 144:43–52
20. Zhou W, Xu K, Ma G, Yang L, Chang X (2016) Effects of particle size ratio on the macro- and microscopic behaviors of binary mixtures at the maximum packing efficiency state. Granul Matter 18(81). https://doi.org/10.1007/s10035-016-0678-1
21. Chang CS, Meidani M (2012) Dominant grains network and behavior of sand-silt mixtures: stress-strain modeling. Int J Numer Anal Methods Géoméch 37:2563–2589
22. Simpson DC, Evans TM (2016) Behavioral thresholds in mixtures of sand and kaolinite clay. J Geotech Geoenviron Eng 142(2):04015073
23. Ahmed I, Lovell CW (1993) Rubber soils as lightweight geomaterials. Transp Res Rec 1422:61–70
24. Bosscher PJ, Edil TB, Kuraoka S (1997) Design of highway embankments using tire chips. J Geotech Geoenviron Eng 123:295–304
25. Iolli S, Modoni G, Chiaro G, Salvatore E (2015) Predictive correlations for the compaction of clean sands. Transp. Geotech 4:38–49
26. Chiaro G, Indraratna B, Tasalloti SMA, Rujikiatkamjorn C (2015) Optimisation of coal wash–slag blend as a structural fill. Ground Improv 168(1):33–44
27. Indraratna B, Qi Y, Ngo TN, Rujikiatkamjorn C, Neville T, Ferreira BS, Shahkolahi A (2019) Use of geogrids and recycled rubber in railroad infrastructure for enhanced performance. Geosciences 9:30
28. Chiaro G, Palermo A, Granello G, Tasalloti A, Stratford C, Banasiak LJ (2019) Eco-rubber seismic-isolation foundation systems: a cost-effective way to build resilience. In: Proceedings of the 2019 Pacific conference earthquake engineering, Auckland, New Zealand, pp 8
29. Hernandez E, Palermo A, Granello G, Chiaro G, Banasiak L (2020) Eco-rubber seismic isolation foundation systems, a sustainable solution for New Zealand context. Struct Eng Int 30(2):192–200
30. Tsang HH, Tran DP, Hung WY, Pitilakis K, Gad EF (2021) Performance of geotechnical seismic isolation system using rubber-soil mixtures in centrifuge testing. Earthq Eng Struct Dyn 50(5):1271–1289
31. Senetakis K, Anastasiadis A, Pitilakis K (2012) Dynamic properties of dry sand/rubber (RSM) and gravel/rubber (GRM) mixtures in a wide range of shearing strain amplitudes. Soil Dyn Earthq Eng 33:38–53
32. O'Sullivan C (2011) Particulate discrete element modelling: a geomechanics perspective. Taylor & Francis, London
33. Itasca Consulting Group (2018) Particle flow code in three dimensions—PFC3D
34. Cundall PA, Strack ODL (1979) Discrete numerical model for granular assemblies. Géotechnique 29(1):47–65
35. Asadi M, Mahboubi A, Thoeni K (2018) Discrete modeling of sand–tire mixture considering grain-scale deformability. Granul Matter 20:1−13
36. Ren ZL, Cheng YP, Xu X (2020) A DEM method for simulating rubber tyres. Géotechnique Lett 10:73–79
37. Mindlin RD, Deresiewicz H (1953) Elastic spheres in contact under varying oblique forces. J Appl Mech 20:327–344
38. Potyondy DO, Cundall PA (2004) A bonded-particle model for rock. Int J Rock Mech Min Sci 41:1329–1364

39. Garcia FE, Bray JD (2019) Modeling the shear response of granular materials with discrete element assemblages of sphere-clusters. Comput Geotech 106:99–107
40. Zhang M, Yang Y, Zhang H, Yu HS (2019) DEM and experimental study of bi-directional simple shear. Granular Matter 21:1–13
41. Rothenburg L, Bathurst RJ (1989) Analytical study of induced anisotropy in idealized granular materials. Géotechnique 39:601–614
42. Radjai F, Wolf DE, Jean M, Moreau JJ (1998) Bimodal character of stress transmission in granular packings. Phys Rev Lett 80:61–64
43. Guo N, Zhao J (2013) The signature of shear-induced anisotropy in granular media. Comput Geotech 47:1–15
44. Huang X, Hanley KJ, O'Sullivan C, Kwok CY (2014) Exploring the influence of interparticle friction on critical state behaviour using DEM. Int J Numer Anal Methods Geomech 38:1276–1297
45. Lopera Perez JC, Kwok CY, Senetakis K (2017) Investigation of the micro-mechanics of sand–rubber mixtures at very small strains. Geosynth Int 24:30–44

Construction of Reinforced Soil Structure in a Lateritic Soil—A Case Study of the Expansion of the Runway at Mangalore International Airport

Minimol Korulla and Beena K. S.

Abstract Mangalore International Airport is one of the few airports in India with a tabletop runway. A tabletop runway is basically a runway located on the top of a hill with ends adjacent to steep slopes. Such airports are challenging not only from the take-off and landing points of view but also from the structural perspective. There is a need for massive cutting and filling operations to construct the runway on mountainous terrain. Stabilization of the valley slope requires slope stabilization measures in the form of retaining structures with a greater height. However, the selection of a suitable retention system for the runway expansion was quite challenging due to the vast stretches of laterite-capped hillocks which have an excellent geotechnical property in dry state and inferior properties in wet conditions. Moreover, the proposed retention system should act as a firm foundation and be able to carry the loads of aircraft, other vehicles, and seismic loading. The total height of the retention system on the runway valley side ranged between 40 and 50 m. The fill material consists of a reinforced soil structure at the top (of max. 17 m high) and bottom (of max. 13 m high) and filled slope in between. The filled slope was protected with berms, erosion control measures, and reinforced with geogrids. Special attention was given to the drainage system so as to reduce the pore water pressure behind the reinforced soil structure and to channelize runoff from the slope through appropriate drains. This paper comprises the system details of a steep reinforced soil structure along with drainage measures and erosion control measures adopted for slope stabilization for runway expansion, the design considerations, and corresponding stability checks for safety with respect to the reinforced soil structure and drainage, post-movement observations, forensic analysis of the reinforced soil structure, and redesign proposal for rehabilitation, current status of the project, and conclusions.

Keywords Reinforced soil structure · Laterite · Drainage · Slope stabilization

M. Korulla (✉) · Beena K. S.
Cochin University of Science and Technology, Kochi, Kerala, India
e-mail: m.korulla@maccaferri.com

Beena K. S.
e-mail: beena@cusat.ac.in

H. Hazarika et al. (eds.), *Multidisciplinary Approach for Disaster Resilience and Sustainable Technologies*, Lecture Notes in Civil Engineering 642,
https://doi.org/10.1007/978-981-96-8618-6_11

1 Introduction

Mangalore airport is one of the few airports in India which is having a tabletop runway (refer Fig.1 for Google Earth image) A tabletop runway is a runway located on the top of a hill with ends adjacent to steep slopes. Such airports are going to be challenging not only from the take-off and landing points of view but also from the structural and slope stability perspective. The nature of the Mangalore airport makes it imperative to analyze the slope stability and to make the appropriate design decisions to counter potential problems. The major structural aspect associated with the Mangalore airport is the stability of the runway and the stability of adjacent side slopes enclosing the runway due to the load of the planes, and potential seepage for forces with uncertain laterite behavior in the presence of moisture.

After the tragic plane crash in 2010, it was decided to extend the runway to 9000 feet (2740 m). The existing runway was found to be inadequate for landing large aircraft. It was suggested that runway extension would enable aircraft of all sizes from home and abroad to arrive directly to Mangalore.

Since the apron is currently attached to the runway's edge, aircraft movements on it must stop upon landing and only begin once reaching the apron. In order to allow aircraft on the apron to proceed toward the runway and hold, a new 2400-m taxiway that runs parallel to the runway and joins it before the landing threshold is planned. This will save the aircraft 10–15 min of flight time.

Fig. 1 Google Earth image of Mangalore International Airport

Fig. 2 Project site before construction—overview of site, locating the reinforced soil structures and existing reinforced concrete wall

2 Location

The project site is located at Bajpe, a locality in Mangalore city of Dakshina Kannada District in the state of Karnataka. The initial project site photographs are shown in Fig. 2.

3 Generalized Geology and Strata Conditions

Mangalore, also known as the Gateway of Karnataka, is the most important port city in India. It is about 352 km west of Bangalore, the capital city. It is surrounded by the Western Ghats hillock in the east, Arabian Sea in the west, Kulur River in the North, and Netravathi River in the south. It has an average elevation of 22 m above mean sea level. The city's topography consists of hills in the Western Ghats and plain up to 30 km from the coast. Preliminary geological studies revealed that the Western Ghats' hills are surrounded by hard laterite. Due to continuous seafloor spreading, the region is subjected to moderate seismic activity from time to time and is categorized under seismic zone-III.

The major rock types are (1) Migmatites, Granites, and gneisses of Peninsular Gneissic complex, (2) high grade schist's of Sargur group, and (3) Meta Volcano-Sedimentary sequence of Bababudan Group. Laterites are present in the vast stretches and are formed mainly in the alteration of the underlying rocks.

The airport land is structurally stable with laterite cover at the top and underlain by very hard and fresh weathered granitic rocks. Major part of the area is covered by the red laterite, and in some portions, pale to bright red is also presented. The valley slopes are covered with loam soil, and the hillside slopes are covered with coarser soil.

The district experiences 3789.9 mm of annual rainfall, most of which falls between June and September during the southwest monsoon season. Over 85% of the yearly precipitation falls during the monsoon season. Approximately 8% of the yearly rainfall occurs during the post-monsoon season, which spans from December to March.

The groundwater occurrence for the project section is mainly in laterites and Granite gneiss formations. In laterites, the inherent porosity, jointed nature, and fractures control water-bearing capacity. In Granites and Granite gneiss formation, groundwater occurs under unconfined, semi-confined, and confined conditions in weathered and fractured zones. Large fluctuation in the water table across the year is observed in Mangalore. The groundwater level is generally shallow in the region, and during summer season, the extraction levels increase leading to considerable decline in groundwater levels.

4 Geotechnical Investigation

Geotechnical investigations were conducted to assess the stratigraphy and to evaluate the subsoil properties below the ground level. Various boreholes at desired locations were proposed, and investigations were carried out in accordance with the standards up to a depth of 6–13 m, to obtain appropriate geotechnical design data. A very dense laterite (Reddish/Brownish/Yellowish dense silty sand with gravel) was encountered at 8–10 m depth followed by weathered rock. Standard penetration tests were conducted at regular intervals in accordance with IS 2131-1981. Several laboratory investigations, such as Atterberg limits, grain size distribution, specific gravity, direct shear tests, triaxial tests, density, UCC tests, were conducted as per the Indian standards.

4.1 Characteristics of Lateritic Soil

Lateritic soils are formed in the regions of high rainfall and high temperature where the climate is humid. These soils are rich in iron and aluminum and formed due to prolonged rock weathering under highly oxidizing and leaching conditions. Lateritic soils contain clay minerals, and they tend to be silica poor and silica leached out by water passing through the soil. Laterite soil is mainly found in Western Ghats and its foothills. These soils expose excellent geotechnical property under dry conditions and poor properties under saturated conditions. Due to an excessive buildup of water in the lateritic soil's pores, the sliding will take place at the interface of impermeable and permeable layers and may contribute to the movement of loose overburn material which in turn causes landslide.

5 Construction of Leveled Landing Platform for Taxiway and Runway with Retention Measures and Other Measures—Initial Scheme

Construction of a leveled platform for taxiway and runway extension required huge filling operations. For runway retention, retaining structures of greater heights are required in the valley side to stabilize the fill slopes. The presence of laterite made the selection of a suitable retention system more challenging. Therefore, a retention system in the form of reinforced soil system which can withstand aircraft loading, other vehicle loading, and seismic loading (zone-III) was selected.

5.1 Proposed System—Initial Scheme

The major underlying aspect surrounding this project site is the stability of the runway side slopes. The stability of a slope is the ability of the slope to withstand the self-weight and external forces simultaneously. In this case, the major loads that the slope has to withstand are the traffic loads due to the aircraft and other vehicles involved in the runway for transportation. Few of the stretches were having space constraints and are required to be stabilized against challenging driving forces with a retention system. For safety, the design of runway and the associated retention/protection work at this location must aim at designs robust enough to (a) account for adverse geological and geotechnical features, and large fluctuations in the water table across the year (b) withstand seismic forces and rain-induced excess hydrostatic pressures within the slope mass. To retain the huge amount of fill at the valley side for the expansion of runway, retention measures of height varying from 40 to 50 m approximately were required to be constructed as shown in Fig. 3. Based on the detailed study of the site conditions and after accounting for the factors discussed above, with due diligence, it was decided to provide a flexible and monolithic reinforced soil system [8, 9, 10] with free draining facia, i.e., gabion facia at the top (max. 17 m high) and bottom of the slope (max. 13 m high) as shown in Fig. 3 (as per MoRT&H Section 3100 [1], BS 8006 Part-1 [2], and FHWA-NHI-10-024 [3]). The reinforced soil structure was properly designed for the anticipated traffic and seismic load for retention purpose.

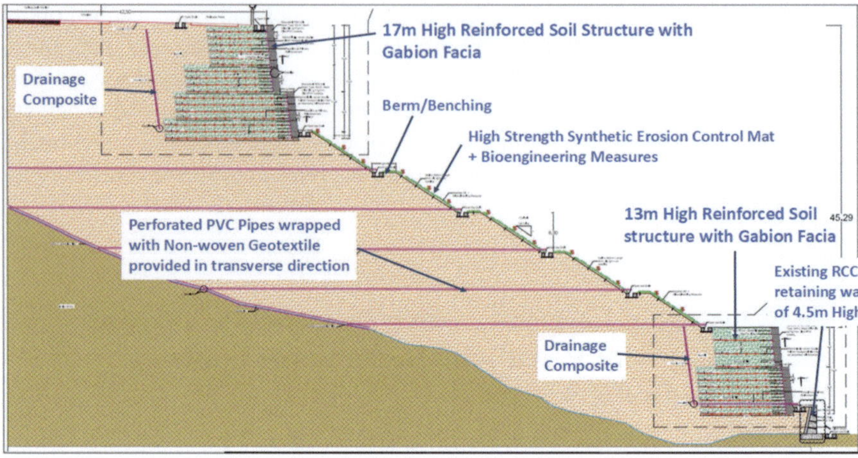

Fig. 3 Proposed solution—retention measures + erosion control measures + drainage measures

There is an existing reinforced concrete retaining wall of 4.5 m high standing at the bottom of the slope (refer Fig. 3).

The filled-up slope between the two reinforced soil structures was provided with berms of 3 m width in between. A combination of synthetic erosion control mat and bioengineering measures was proposed to be provided as an erosion control measure to protect the slope surface between two reinforced soil structures (as per MoRT&H Section 700 [1], IRC 56 [4]).

The design gave special attention to the efficacy of the drainage system to reduce the pore water pressure behind the reinforced soil structure and to channelize the surface and subsurface runoff from the slope through appropriate drains as per IRC SP 42 [5]. Drainage behind the two reinforced soil structures was addressed by providing the drainage composite as a chimney drain and was proposed to be connected with longitudinal and transverse perforated PVC pipes wrapped with non-woven geotextile. The perforated PVC pies were further connected externally to the open toe drain as shown in Fig. 3. The drainage between in situ soil (laterite) and backfill is ensured by providing 300 mm aggregate filter material wrapped with non-woven geotextile. The drainage of the filled-up slope between the two reinforced soil structures was addressed by providing perforated PVC pipes wrapped with non-woven geotextile in traverse direction and was externally connected to the open toe drain.

6 Analysis and Design of Various Measures

The analysis and design of the proposed solution scheme include (a) analysis and design of reinforced soil structure using limit equilibrium method with respect to global stability check of entire slope of 40–50 m in static and seismic conditions, wall safety, internal safety, and bearing capacity checks for both top and bottom reinforced soil structures in static and seismic conditions; (ii) drainage safety checks (with regard to permeability)—design of drainage composite as chimney drain behind reinforced soil structure to dissipate the pore water pressure developed in backfill soil.

6.1 Analysis and Design of Initially Proposed Reinforced Soil Structure

The analysis and design of reinforced soil structures includes global stability checks, external stability checks, and internal stability checks. External stability is evaluated by assuming the reinforced soil zone as a coherent mass with lateral earth pressure acting on the back side of the block. Sliding, overturning, and bearing or tilt are the three external failure modes. Internal stability mode involves the failure of reinforcement and depends on mainly three factors, i.e., tensile resistance of reinforcement against rupture, soil-reinforcement interaction for pull out, and internal sliding.

Based on the geometry, existing ground level and the proposed runway levels, critical design sections (17 m high reinforced soil structure at top and 13 m high reinforced soil structure at bottom) were chosen. Aircraft loading of 145 kPa was considered in addition to the traffic load of 24 kPa. Seismic analysis was also done considering horizontal seismic coefficient 0.08 as per IS 1893 [6]. Based on the bore log reports shared by client, the following are the parameters considered in design and analysis of reinforced soil structures (Table 1).

The reinforced soil structure design and analysis were carried out by following British Standard (BS 8006 [2]) for static load case and AASHTO Standards [7] (using FHWA-NHI-10-024 [3]) for the seismic load case.

All stability checks were performed using MacStARS, Manufacturer's (Maccaferri Environmental Solutions Pvt. Ltd.) inbuilt slope stability software (refer Fig. 4). Reinforced soil structures were analyzed for ultimate limit state under static case and serviceability limit state under seismic case. The program allows the user to conduct stability checks using the limit equilibrium method even considering unreinforced slopes and without support works. Tables 2 and 3 indicate the various stability checks analyzed for reinforced soil structure.

Table 1 Geotechnical parameters considered for in situ strata, structural fill, and backfill soil for the design and analysis

Property	In situ strata	Structural fill	Backfill
Cohesion (kPa)	18	10	10
Angle of internal friction (degree)	31	31.5	31.5
Bulk unit weight (kN/m^3)	19.3	18.7	18.7

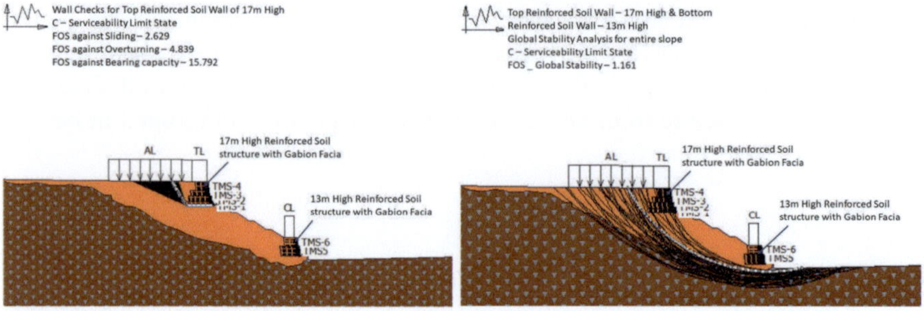

Fig. 4 Snapshots of design and analysis of reinforced soil structures

Table 2 Summarized factors of safety (FoS) for soil reinforced structure (static case design as per BS 8006-1:2010 [2])

Property	Target FoS	Achieved FoS for top structure	Achieved FoS for bottom structure
Global stability for entire slope	> 1.00	1.273	
Global stability	> 1.00	1.343	1.166
Sliding stability	> 1.00	2.087	1.814
Overturning	> 1.00	6.968	3.442
Bearing capacity	> 1.00	6.788	3.006
Internal stability	> 1.00	1.359	1.174

Table 3 Summarized FoS for soil reinforced structure (seismic case design as per Annexure F of FHWA-NHI-10-024 [3])

Property	Target FOS	Achieved FoS for top structure	Achieved FoS for bottom structure
Global stability for entire slope	> 1.00	1.161	
Global stability	> 1.00	1.463	1.276
Sliding stability	> 1.125	2.629	1.489
Overturning	> 1.5	4.839	1.608
Bearing capacity	> 1.875	15.792	3.776
Internal stability	> 1.125	2.042	1.308

From the results, the FoS obtained from the analysis is greater than the target FoS. Hence, the proposed structure will be safe and stable.

Even though from the above analysis, all the checks are safe but, in the actual field, while constructing the bottom reinforced soil structure of 13 m high, there were some cracks developed in the backfill of RS wall. The details of post-movement observations and forensic analysis are given in clause no.7.

6.2 Design of Drainage Composite as Chimney Drain Behind Reinforced Soil Structure to Dissipate the Pore Water Pressure Developed in Backfill Soil

The input soil sample details considered for drainage composite design are given below (Table 4).

Horizontal earth pressure on the drainage composite

$$P = K_a * \gamma * H = 0.314 * 20.7 * 15 = 97.497 \text{ kPa} \tag{1}$$

Step-1 Calculation of in-plane flow rate of drainage composite from manufacturer data sheet

Reduction factors for drainage composite (for design life ≥ 100 years).

- For intrusion of filter geotextiles into the draining core = $RF_{ID} = 1.25$.
- For creep = $RF_{cr} = 1.078$.
- For chemical degradation = $RF_{CD} = 1$.

Table 4 Input details for drainage composite design

S. No	Property	Value
1	Maximum height of backfill behind drainage composite (H) (m)	15
2	Rainfall intensity (mm/hour)	80
3	Particle size distribution	
	D10 (mm)	0.015
	D15 (mm)	0.018
	D30 (mm)	0.021
	D60 (mm)	0.45
	D85 (mm)	1.75
4	Atterberg limits	
	Liquid limit (%)	38.87
	Plastic limit (%)	19.47
	Plasticity index (%)	19.4
5	Coefficient of permeability, k (mm/sec)	3.8×10^{-03}
6	Saturated unit weight (kN/m^3)	20.7
7	Angle of internal friction (degree)	31.5
8	Coefficient of active earth pressure (k_a)	0.314

- For biological degradation = $RF_{BD} = 1$.

$$\begin{aligned}\text{Total reduction factor} &= RF_{ID} \times RF_{cr} \times RF_{CD} \times RF_{BD} \\ &= 1.25 \times 1.078 \times 1 \times 1 = 1.3475 \end{aligned} \tag{2}$$

RF_{Ir} = Empirical factor for contact conditions different from project conditions = 0.5

Drainage composite in-plane flow rate from manufacturer data sheet at a hydraulic gradient (i = 45-degree slope), pressure 97.497 kPa = 0.55 l/m^{-s}.

$$\begin{aligned}50\,\%\ \text{Reduction in flow} &= 0.55\,l/m^{-s} \times RF_{Ir} \\ &= 0.55 \times 0.5 = 0.275\,l/m^{-s}\end{aligned} \tag{3}$$

$$\begin{aligned}&\text{Long-term flow rate for drainage composite} \\ &\quad = \frac{\text{short term flow rate obtained from laboratory tests}}{\text{Total Reduction Factor, RF}} \\ &\quad = \frac{0.275}{1.3475} = 0.204\ l/m^{-s}\end{aligned} \tag{4}$$

Step-2 Calculation of flow rate of drainage composite from field

Flow rate of drainage composite at actual field conditions was determined from slope stability analysis software (slide software), and the snapshot of the analysis is indicated below.

Actual flow happening in field from Fig. 5, Q = 0.0002083 cumec = 0.2083 l/ m^{-s}.

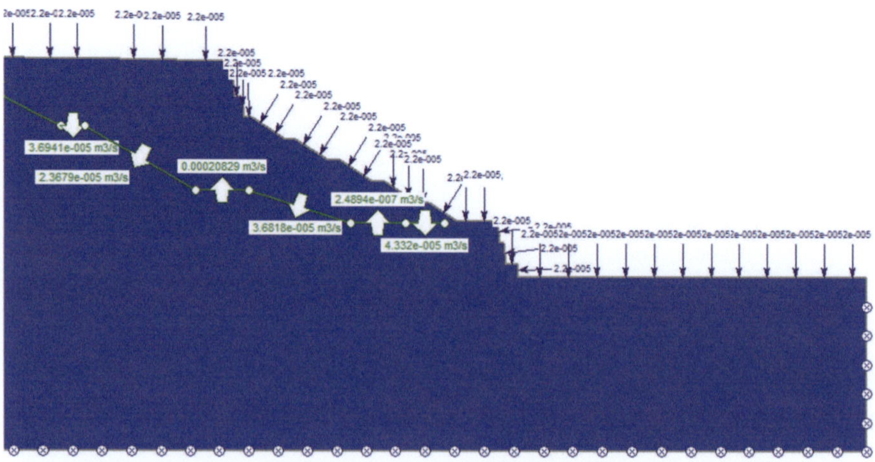

Fig. 5 Slide analysis indicating the actual flow happening in the field

Step-3 Factor of safety considering flow rate conditions for drainage composite

$$= \frac{\text{Long term flow rate for drainage compsote}}{\text{Total Discharge form backfill soil which is actually happening in the field}}$$
$$= \frac{0.204}{0.2083} = 0.979 \leq 1, \text{ unsafe} \quad (5)$$

Hence, the combination of two drainage composites was proposed to be placed as a chimney drain behind the reinforced soil structure to meet the FoS requirements.

$$\text{Long-term flow rate for combination of two drainage composites}$$
$$= 2 * \frac{0.275}{1.3475} = 0.408\,\text{l/m}^{-s} \quad (6)$$

FoS considering the flow rate conditions for the combination of two drainage composites

$$= \frac{0.408}{0.2083} = 1.96 \geq 1, \quad \text{safe} \quad (7)$$

Hence, two layers of drainage composite are suitable considering flow rate criteria.

Step-4 Design of geotextile component of drainage composite

The geotextile must have sufficient physical and mechanical properties for proper functioning of drainage composite as a drainage layer. The geotextile shall have proper permittivity and Apparent Opening Size (AOS) to suit the site-specific soil requirements.

Geotextile filter properties from manufacturer technical data sheet

Apparent Opening Size (mm) = 0.22.
Permittivity of Geotextile (per sec) = 0.5.
Permeability (mm/s) = 0.5.

Retention criteria

$$\text{AOS or } O_{95} < B * D_{85} \quad (8)$$

where

AOS is Apparent Opening Size
O_{95} is opening size in the geotextile for which 95% are smaller
B is Dimensionless coefficient
D_{85} is Soil particles size for which 85% are smaller.

For sands, gravelly sands, clayey sands (with less than 50% passing the 0.075 mm sieves), B is a uniformity coefficient function, C_u.

Therefore, for

$$C_u \leq 2 \text{ or } \geq 8 \;\; B = 1 \tag{9}$$

$$2 \leq C_u \leq 4 \;\; B = 0.5\,C_u \tag{10}$$

$$4 < C_u < 8 \;\; B = 8/C_u \tag{11}$$

For silts and clays (with more than 50% passing the 0.075 mm sieve), B is a function of type of geotextile:

$$\text{For woven} \quad B = 1; \; O_{95} \leq D_{85} \tag{12}$$

$$\text{For nonwovens} \quad B = 1.8; \; O_{95} \leq 1.8\,D_{85} \tag{13}$$

$$\text{And for both} \quad \text{AOS or } O_{95} \leq 0.3\,\text{mm} \tag{14}$$

$$B = 1.8 \times D_{85} = 1.8 * 1.75 = 3.15 \tag{15}$$

$$\text{AOS or } O_{95} < B * D_{85} = 0.22 < 3.15 = 0.22 < 3.15, \;\text{safe} \tag{16}$$

Clogging Resistance Check

At the same time, the pore size geotextile must be large enough so that the fine backfill particle may not clog the geotextile pores.

$$O_{95} \geq BD_{85}(\text{soil}) \quad \text{for } C_u < 3 \tag{17}$$

$$O_{95} \geq 3D_{15}(\text{soil}) \quad \text{for } C_u > 3 \tag{18}$$

$$C_u = \frac{D_{60}}{D_{10}} = \frac{0.45}{0.015} = 30 \tag{19}$$

$$0.22 \geq 3 * 0.018 = 0.22 \geq 0.054, \;\text{safe} \tag{20}$$

Permeability Check

The coefficient of permeability of filter geotextile must be 10 times larger than the permeability of backfill soil.

$$k_{\text{geotextile}} > 10\,k_{\text{soil}} \tag{21}$$

$$0.5 > 10 * 3.8 \times 10 - 03 = 0.5 > 0.038, \text{ safe} \tag{22}$$

The suggested geotextile is appropriate from the filtration point of view as it satisfies all three requirements of retention criteria, permeability requirement, and clogging criteria.

7 Post-movement Observations and Forensic Analysis and Redesign Proposal for Rehabilitation

One of the ongoing issues that has been encountered in this process was the development of cracks in the backfill of the bottom reinforced soil structure during construction. The major factor attributing to this problem was the shorter length of the primary reinforcement at the top tier of bottom reinforced soil structure (13 m high), than the required length. Therefore, slope stability analysis was performed using slide software to check the stability requirements. The properties considered for the analysis using slide software are the same as given in clause no 6.1. The snapshots of the analysis of reinforced soil structures are shown in Fig. 7.

From the analysis, it was found that the primary reinforcement length needs to be increased from 10 to 24 m (max.) at top tier for 17 m high reinforced soil structure and 10–20 m (max.) at top tier for 13 m high reinforced soil structure. Additional geogrid layers were provided in the fill slope between two reinforced soil structures due to the huge amount of fill material involved and to meet the FoS requirement globally and locally.

The filled-up slope between the two reinforced soil structures was protected with stone pitching, and the stone pitching was provided between two double-twisted steel wire mesh panels as shown in. It is recommended that vegetation be placed in the voids left by stone pitching. The whole system is secured with shorter nails of 1.5 m long and 2 m c/c spacing in horizontal and vertical directions to hold the pitching from erosion. Appropriate modifications were made in the drawings as shown in Fig. 6. The results from this analysis help to elucidate this as the most prudent solution for runway expansion.

8 Current Status of the Project and Conclusion

Currently, the project is under progress and various slope stabilization measures are either being placed or in the process of execution. The site photographs are shown in Fig. 8.

One of the significant messages for highway engineers engaged in slope stabilization projects especially in hilly areas is that no matter how rigorous is the investigation and design, project team should always remain vigilant and take pro-active steps to

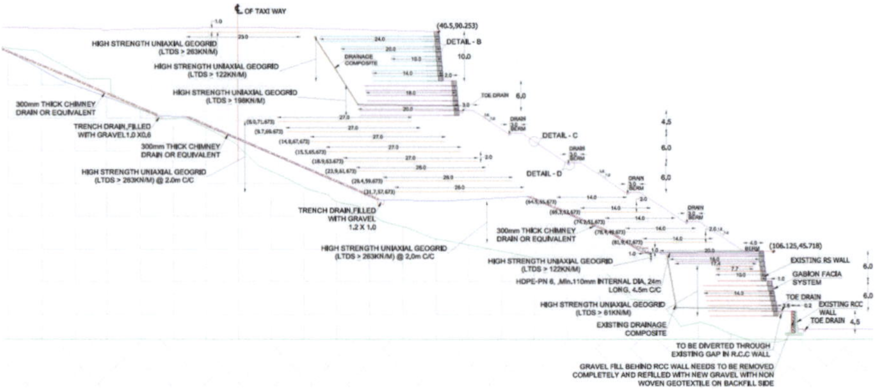

Fig. 6 Redesigning the original reinforced soil structure

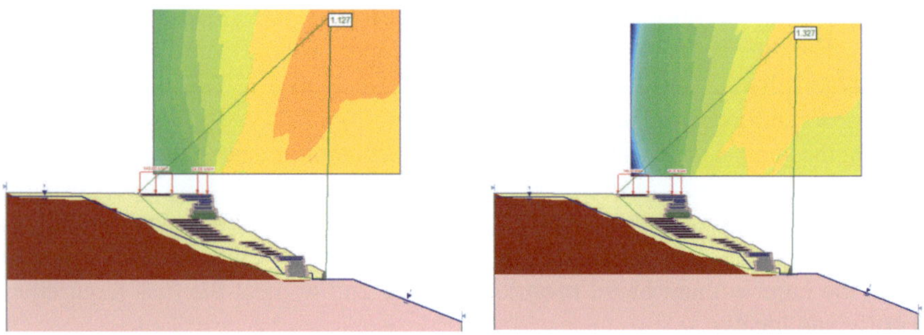

Fig. 7 Software outputs—reinforced soil design and analysis (left—seismic case and right—static case)

Fig. 8 Site photographs showing the various stabilization measures adopted at site

validate the design assumptions through real-time field monitoring. Modifications may be required in the design/proposed scheme at any stage during the execution of the project because of uncertainties, extreme weather events etc.

The general design and analysis methods might not always be adequate for the complete design of the reinforced soil structures. In some situations, additional analysis and treatments could be necessary to ensure durability, serviceability, and safety.

Tall reinforced soil structures (height > 15 m), reinforced soil structures with space constraints, extremely poor ground conditions, extreme environmental conditions, etc., are a few examples of such structures. Reinforced soil structures will be highly complicated and prone to movements or displacements over time when they fall under any of the various combinations of the aforementioned scenarios. Numerical analysis must be carried out in order to check the deformations and to validate the results with those obtained from theoretical standard design techniques.

In tall reinforced soil structures, shear stresses within the soil mass develop and overburden pressure rises with wall height. The wall's face has a propensity to move outward, and this tendency gets stronger the higher the wall gets. Therefore, it is recommended to provide more length for the upper layers of primary reinforcement. This proportion limits the potential for developing tension cracks at the interface and resists lateral loading effects.

References

1. MoRTH (5th revision) Specifications for road and bridge works. Indian Roads Congress, India
2. BS 8006 Part1 (2010) Code of practice for strengthened/reinforced soils and other fills
3. FHWA-NHI-10-024 (2009) Design and construction of mechanically stabilized earth walls and reinforced soil slopes. Federal Highway Administration, Washington, D. C
4. IRC 56 (2011) Recommended for practice for treatment of embankment and roadside slopes for erosion control. Indian Roads Congress, India
5. IRC SP 42 (2014) Guidelines for road drainage. Indian Roads Congress, India
6. IS 1893 Part1 (2002) Recommendations for earthquake resistant design of structures. Bureau of Indian Standards, India
7. AASHTO, LRFD (2012) Bridge design specifications. American Association of State Highway and Transportation Officials, Washington, DC
8. Korulla M, Budhbhatti R, Meenu PS, Kamal V (2019) Sustainable landslide mitigation measures-demonstration through a live case study in Uttarakhand. In: Frontiers in geotechnical engineering, developments in geotechnical engineering, chapter-19. Springer Nature Singapore Pte Ltd., pp 395–415
9. Korulla M, Mourya PK, Khan S, Kachhal A (2018) Landslide and rockfall mitigation works in Uttarakhand—case study. In: ISGTI conference, India
10. Korulla M, Chaurasia S (2018) Sustainable landslide mitigation measures—a case study on Lambagarh. In: ISGTI conference, India

Sheet Pile with Drainage Ability to Reduce Liquefaction-Induced Embankment Settlement

Mitsu Okamura, Mizuki Suyama, and Kohei Ono

Abstract In urban areas, sheet piling is preferably employed in the remediation program for river levees against liquefaction-induced failure. It is often the case that sheet piling is not sufficient to maintain levees on liquefied foundation soil high enough to avoid overflooding. Sheet piles with additional members with drainage ability have been developed and used in practice, which is expected to have the combined effects of providing the lateral confinement by the sheet pile and facilitating dissipation of excess pore pressure from liquefiable soils below the levee. This paper describes a series of centrifuge tests to assess seismic settlement of levee and excess pore pressure in foundation soil enclosed with sheet piles with and without drainage ability. Effects of drainage ability were studied for wide range of permeability of liquefiable foundation soil. Experimental data were used to verify the validity of the pore pressure prediction procedure for soil below levees. It was confirmed that excess pore pressure in the sand bed (Δu) is quite sensitive in a small range of soil permeability above which Δu is high and practically constant irrespective of the sand permeability. The drainage ability of the sheet pile has a significant effect to decrease the permeability of the range, especially for smaller levees.

Keywords Liquefaction · Embankment · Sheet pile · Drain · Centrifuge test

M. Okamura (✉)
Waseda University, Tokyo, Japan
e-mail: m.okamura@waseda.jp

M. Suyama
Nippon Koei, Tokyo, Japan

K. Ono
Ehime University, Matsuyama, Japan

H. Hazarika et al. (eds.), *Multidisciplinary Approach for Disaster Resilience and Sustainable Technologies*, Lecture Notes in Civil Engineering 642,
https://doi.org/10.1007/978-981-96-8618-6_12

1 Introduction

After the 1995 Hyogoken-Nanbu earthquake which made a part of the Yodo River levee slump and the hinterland in Osaka city in real danger of flooding, a remedial countermeasure project against liquefaction-induced failure of vulnerable levees in high-risk areas has been advancing in Japan. Early studies using shaking table tests and centrifuge tests on embankments resting on loose saturated sand [1, 2] revealed that the major cause of the embankment settlement was the lateral deformation of liquefied soil away from the embankment centerline. Countermeasures in the program aim at mitigating crest settlements by providing containment for the deformation of the liquefiable foundation soils by forming remediated stiff zones in liquefiable soil layer under levee toes [3–5]. The remedial measures used in the practice include sheet pile enclosures, solidification by deep mixing, densification by sand compaction piles, and providing drainage ability by gravel drains. They are implemented in the free fields just outside the toes of levee.

In urban areas, the sheet piling is preferably employed especially for locations where construction space is limited on the hinterland side of levees. It is often the case that sheet piling is not sufficient to maintain levees on liquefied foundation soil high enough to avoid overflooding. Sheet piles with additional members with drainage ability have been developed (Fig. 1) and used in practice, which is expected to have the combined effects of providing the lateral deformation confinement by the sheet pile and facilitating dissipation of the excess pore pressure from soils below the levee. However, research works on effects of the sheet pile with drainage ability is very limited. More studies are apparently needed to better understand the effects.

Effects of drainage are measured by dissipation of the excess pore pressures in soils. The currently employed vertical drain design uses the procedure originally developed by Seed and Booker [6]. They demonstrated that the excess pore pressure in sand around a drain with level ground surface can be analyzed using the diffusion

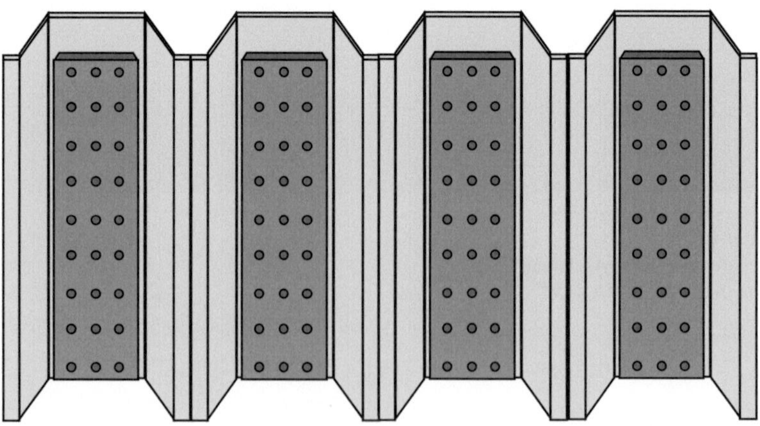

Fig. 1 Schematic illustration of sheet pile with members of drainage ability

equation incorporated with generation of the excess pore pressures owing to the earthquake cyclic shearing. For cases of two-dimensional diffusion considered in this study with anisotropy in permeability taken into account, the excess pore pressure (Δu) can be expressed as,

$$\frac{\partial \Delta u}{\partial t} = \frac{\partial \Delta u_g}{\partial t} + \frac{1}{m_v \gamma_w}\left(k_x \frac{\partial^2 \Delta u}{\partial x^2} + k_z \frac{\partial^2 \Delta u}{\partial z^2}\right) \tag{1}$$

where $\partial \Delta u_g / \partial t$ stands for the generation rate of the excess pore pressure due to the earthquake cyclic shearing in undrained conditions, determined by the cyclic shear stress time history induced by an earthquake and the liquefaction resistance of sand. The second term on the right-hand side expresses the dissipation of the excess pore pressure in the horizontal direction (x) towards drainage member of the sheet pile as well as the vertical direction (z) with dominant soil parameters, the coefficient of permeability in vertical and horizontal directions, k_x and k_z, and the coefficient of volumetric compressibility, m_v. The latter coefficient is defined as $m_v = \varepsilon_v / \Delta u$ where ε_v is volumetric strain produced by dissipation of Δu. Hence, m_v is dependent on the stress level [7].

Regarding the soil parameters, Seed and Booker suggested to use a constant value of m_v irrespective of the excess pore pressure ratios and confining pressures, because the most influential soil parameter in the diffusion process would be the soil permeability, which potentially varies in a much wider range of values than m_v. This assumption that m_v is a stress-independent constant has been handed over to design procedures in the current practice [8]. However, depending on the confining stress level, the variation of m_v can be more than several times, and thus, assuming m_v to be constant regardless of stress level or depth is not a reasonable approach. Minaka et al. [9] clearly demonstrated that the assumption of constant m_v estimated the excess pore pressure on the unsafe side.

This paper describes a series of centrifuge tests to assess seismic settlement of levee and excess pore pressures in foundation soil enclosed with sheet piles with and without drainage ability. Effects of drainage ability were studied using models with permeability of liquefiable foundation sand varied in a wider range. Experimental data were used to verify the validity of the pore pressure prediction procedure for soil below levees. The pore pressure model proposed by Okamura [10] is incorporated in the generation term to simulate the irregular input acceleration time history observed in the tests. Stress-level dependent and anisotropic soil parameters determined from the laboratory tests were employed in the analysis.

2 Centrifuge Test

2.1 Model Preparation and Test Conditions

A series of centrifuge tests was carried out on model levees resting on a liquefiable saturated sand bed with sheet piles inserted at both toes, as illustrated in Fig. 2a. Toyoura sand deposited at a relative density Dr = 50% was used in all eight centrifuge tests described in this paper (Table 1). It is a fine sand of $D_{50} = 0.15$ mm and a permeability (for the sand at Dr = 50%, tested at 1 g using water as pore fluid in the direction perpendicular to the sand deposition plane) of $k_z = 2.0 \times 10^{-4}$ m/s. The sand shows anisotropy in permeability coefficient; the coefficient of horizontal permeability, k_x, (parallel to the sand deposition plane) was observed in the laboratory tests to be eight times higher than that vertical (k_z). The specific gravity and minimum and maximum void ratio of the sand are $G_s = 2.64$, $e_{\min} = 0.609$, and $e_{\max} = 0.973$, respectively.

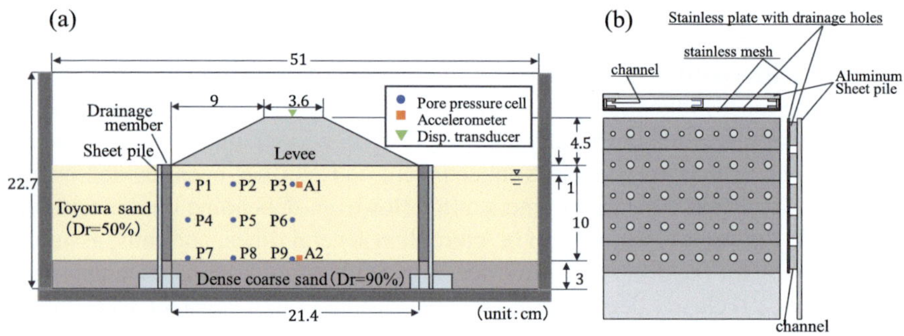

Fig. 2 Centrifuge model. **a** Levee remediated by sheet piles with drainage ability, **b** aluminum sheet pile with members of drainage ability. All dimensions are shown in model scale

Table 1 Centrifuge test conditions

Test code	Drainage member	Pore fluid viscosity, ν (cSt.)	Prototype permeability	
			Vertical, k_z (m/s)	Horizontal, k_x (m/s)
SP-1	No	12.5	8.0×10^{-4}	6.4×10^{-3}
SPD-1	Yes			
SP-2	No	25	4.0×10^{-4}	3.2×10^{-3}
SPD-2	Yes			
SP-3	No	50	2.0×10^{-4}	1.6×10^{-3}
SPD-3	Yes			
SP-4	No	200	5.0×10^{-5}	4.0×10^{-4}
SPD-4	Yes			

A pair of sheet piles was set in a model container with its tip rigidly fixed to the container base not to cause neither translation nor rotation. Dry Toyoura sand was rained at a relative density of 50% to the container with internal dimensions of 510 mm wide, 230 mm deep and 120 mm long. The model sheet piles were aluminum plates of 3 mm thick and its flexural rigidity was 4.3 MPa m^4/m in the prototype. The member to facilitate the drainage ability consisted of channels and 0.5 mm thick aluminum plate with a number of small holes to allow water to flow in the channel. The hole was covered with a stainless steel mesh to avoid the sand to flow into the member (Fig. 2b). The members were attached to a side of sheet pile to allow water to flow from sand under the levee. During sample preparation, accelerometers and pore pressure cells were installed at the proper locations and orientations as indicated in Fig. 2a. A laser displacement sensor was set to observe settlement of embankment crest.

The soil used to construct the levees was a mixture of Toyoura sand and Ube keisa #7 at 2:1 by dry weight. The soil was prepared at water content of 10% and compacted in a wooden mold of trapezoidal shape to a dry density of 90% the maximum dry density. The levee in the mold was frozen in the refrigerator, set on the foundation layer in the container box and thawed in the room temperature. On completion of thawing the embankment, the foundation soil was fully saturated with viscous fluid in a vacuum chamber, to a degree of saturation higher than 99.6% which was measured with the method developed by Okamura and Inoue [11]. The model was moved to the geotechnical centrifuge at Ehime University and spun up to the centrifugal acceleration of 50 g. The viscous fluid on the ground and in the levee soil was drained through a stand pipe connected to the bottom drainage holes until the ground water table stabilized at 10 mm below the ground surface. The model was finally shaken horizontally with a mechanical shaker. The acceleration time histories of the event have the basic shape indicated in Fig. 3 with acceleration amplitude of approximately 2.0 m/s^2 and lasted 50 s in the prototype scale. The input motion in terms of shape of the waves and each peak acceleration were replicated quite well throughout the series of tests.

A total of eight centrifuge tests were carried out. Test conditions are summarized in Table 1. This paper focuses on the effect of drainage member on the sheet pile and the foundation sand permeability on mitigation of levee settlement. The coefficient

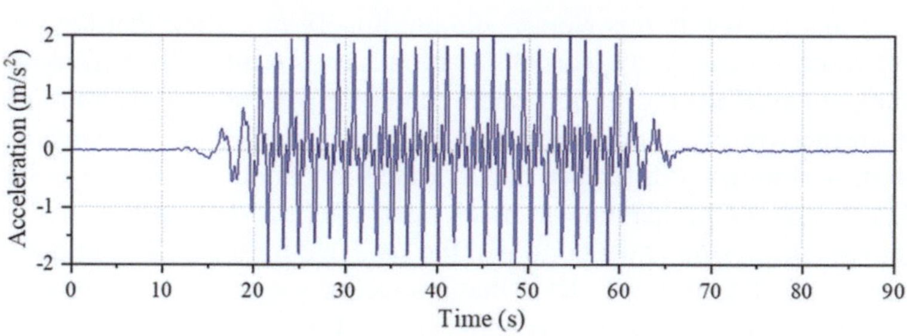

Fig. 3 Time history of input acceleration in prototype scale

of permeability was systematically varied between tests. This was achieved by using viscous fluid with different kinematic viscosity. Metolose solution with viscosity of 12.5, 25, 50, and 200 times the viscosity of water (that is 12.5, 25, 50, and 200 cSt) was used as pore fluid.

One consequence of using a viscous fluid with viscosity of v cSt in a centrifuge test at 50 g to model liquefaction of viscous fluid-saturated prototype soil in the field is that the corresponding prototype permeability being simulated is 50/v times that of the model [12]. In this study, $k_{\text{prototype}}$ ranges from $2.0 \times 10^{-4} \times 50/12.5 = 8.0 \times 10^{-4}$ m/s to $2.0 \times 10^{-4} \times 50/200 = 5.0 \times 10^{-5}$ m/s, which corresponds to the permeabilities of coarse sand and silty sand in the field. Therefore, the eight centrifuge models listed in Table 1 simulated induced liquefaction and levee settlement in the field of the same prototype layer (homogeneous deposit of a uniform Toyoura sand of Dr = 50%) but having different permeabilities remediated with a pair of sheet piles with and without drainage member attached. Hereafter in this paper, all results are presented in prototype units unless otherwise mentioned.

2.2 *Test Results and Discussions*

All eight tests listed in Table 1 simulate the same field situation, differing only in their soil permeability and the drainage member installed to the sheet pile. In this section, comparison between results of tests with different permeability is discussed first, followed by the effects of drainage ability of sheet pile on pore pressure responses.

Excess pore pressure responses and deformation

Time histories of excess pore pressure (Δu) are shown in Fig. 4 for cases of SP-4 and SPD-4 with the lowest permeability during shaking. Dashed lines in the figure represent the initial effective vertical stress (σ'_{v0}) calculated assuming embankment height just above the pore pressure cells as a uniform vertical surcharge. Under the levee toe near the sheet pile (P4 and P7), Δu attained σ'_{v0}, corresponding to the condition of sand liquefaction, in a few cycles after the base shaking started. On the levee centerline (P6 and P9), the excess pore pressures were lower than σ'_{v0} and the acceleration did not attenuate significantly throughout the shaking event, indicating that the soil in this vicinity did not liquefy in a sense that the soil loses its effective stress completely. This observation is consistent with centrifuge tests of embankment model resting on liquefiable sand [4, 5]. Δu at any locations are quite similar irrespective of the drainage ability of the sheet pile. The drainage member did not make significant difference in Δu of the foundation sand between SP-4 and SPD-4, although Δu dissipation was much faster after the shaking ceased.

Figure 5 indicates Δu at the bottom of the sand bed for the models of the highest permeability, SP-1 and SPD-1. Δu in these models were generally lower than those shown in Fig. 4, owing to the higher permeability. Comparison between result from the two tests clearly demonstrates effects of the drainage member. At the very beginning of the shaking Δu in SPD-1 followed those in SP-1 and sharply decreased

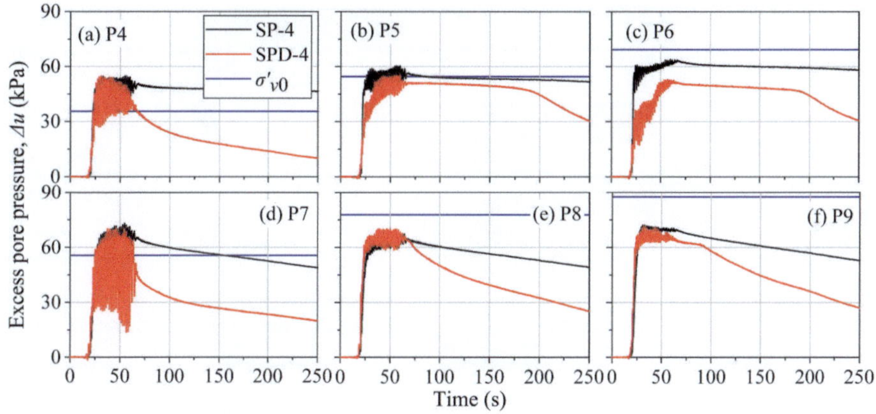

Fig. 4 Excess pore pressures for tests SP-4 and SPD-4 ($\nu = 200$ cSt.)

immediately after Δu reached approximately 10 kPa. The timings when the Δu turn to decrease in SPD-1 were approximately the same at all locations.

Figure 6 shows photos of the models after shaking, in which black lines in the foundation soil are the dyed sand put near the side glass of the container during model preparation so that ground deformation could be easily detected. In the models SP-2, 3, and 4, significant deformation of foundation sand can be observed where sand below the levee was squeezed out towards toes of the levee. The levees of these models were subsided approximately 0.6 m with its base becoming bowl-shaped and stretching laterally. It is of interest to note that the deformation of the model SP-1 is quite small although the permeability of SP-1 is only two times higher than SP-2. This is also the case for model with drainage member. Deformation of model significantly decreased by increasing permeability. Crest settlement dramatically decreased from 0.6 m for SPD-3 to 0.05 m for SPD-2, where the difference in permeability between the two models is only two times.

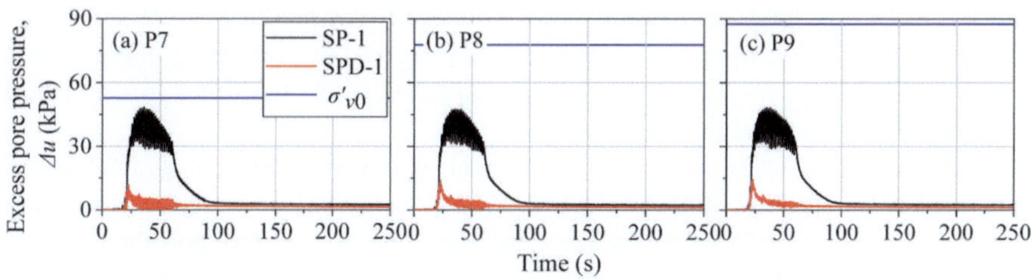

Fig. 5 Excess pore pressures at selected locations for tests SP-1 and SPD-1 ($\nu = 12.5$ cSt.)

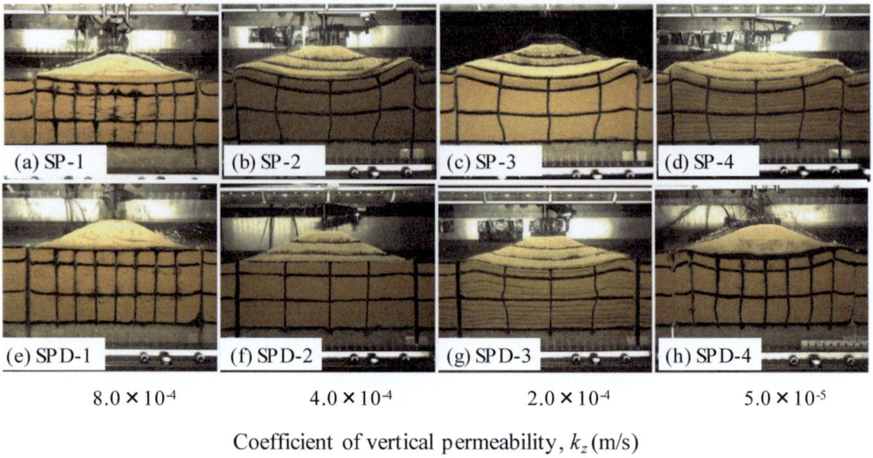

Fig. 6 Photos taken after shaking, showing deformation of the models

Effects of drainage ability

Overall picture of the effects of permeability on the excess pore pressure in the foundation is demonstrated in Fig. 7, in which maximum excess pore pressure ratio ($r_{u\_max} = \Delta u_{max}/\sigma'_{v0}$) at P9 (bottom on the centerline) are plotted against vertical permeability coefficient (k_z) for all models tested in this study. For models without drainage member (SP), r_{u_max} stayed approximately unity in tests SP-4, 3, and 2, and decreased significantly in SP-1. The pore pressure responses changed from undrained to partial drained at the permeability approximately 4.0×10^{-4} m/s. While for models with drainage ability (SPD), although general trend of r_{u_max} with k_z is quite similar to that without drainage member, the threshold k_z where the responses change from undrained to partially drained is approximately five times smaller as compared to that of models without drainage member (SP). r_{u_max} for SPD-3 ($k_z = 2.0 \times 10^{-4}$ m/s) is clearly lower than the less permeable model SPD-4 and model SP-3 with the same permeability but without drainage member. Moreover, r_{u_max} dropped to lower than 0.2 in SPD-1 indicating that drained condition was practically reached for k_z higher than 8.0×10^{-4} m/s. The partial drained occurred in a relatively narrow range of soil permeability; width of the ranges are approximately ten times in k_z for both models with and without drainage member.

Figure 8 depicts crest settlement at the end of shaking. There is practically no difference in the settlement for tests SP-2, 3, and 4 and SPD-4 where the sand bed is considered to be in undrained condition and the soil permeability and the drainage member did not have a clear effect on the levee settlement. Further increase in the permeability made the pore pressure responses of sand beds partially drained and decreased the settlement significantly. It is worth mentioning that the crest settlement decreased with the permeability more rapidly than the excess pore pressure; the settlement of SPD-2 was negligibly small although r_{u_max} was not that low, approximately

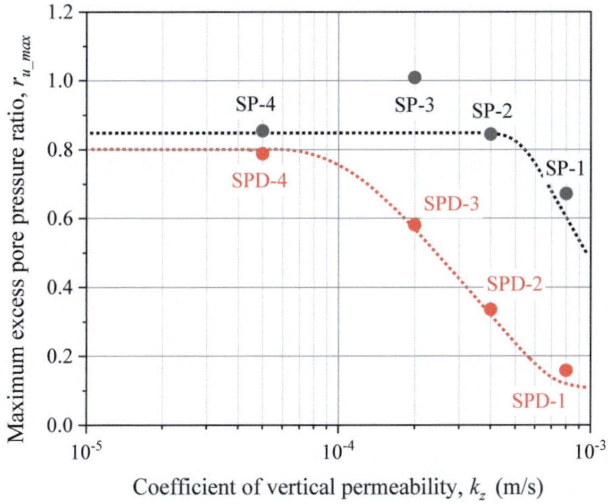

Fig. 7 Evolution of excess pore pressure ratio with sand permeability

0.4. This is because sand does not lose its strength and stiffness significantly until r_u becomes higher than a certain level. It can be concluded that crest settlement changes more sensitively to the soil permeability than Δu.

Figure 9 shows excess pore pressure distributions at the end of shaking along with two representative lines in the sand beds. Figure 9a indicates horizontal distribution on the mid-depth of the sand bed. Δu for less permeable models without the drainage member is more or less uniform. While for models SPD-3 and SPD-4 with the drainage member, Δu near the sheet pile (P4) is lower than the other two locations (P5 and P6), indicating water flow towards sheet pile. Averaged hydraulic gradient between P4 and P6 of these models is 0.25. Vertical distribution of Δu on the levee centerline (P3, P6, and P9) are shown in Fig. 9b. Upward water flow is apparent for

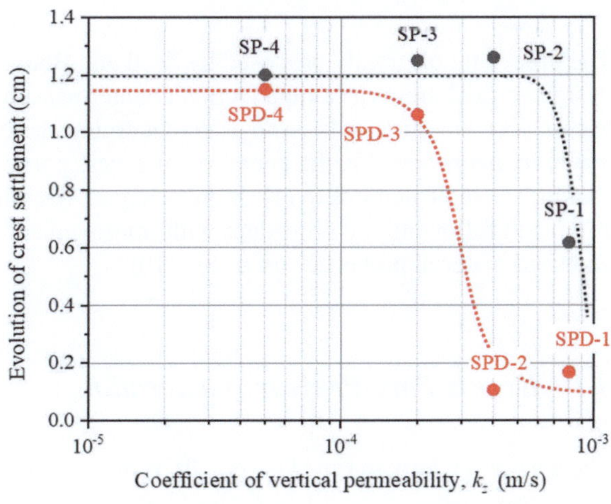

Fig. 8 Evolution of crest settlement with sand permeability

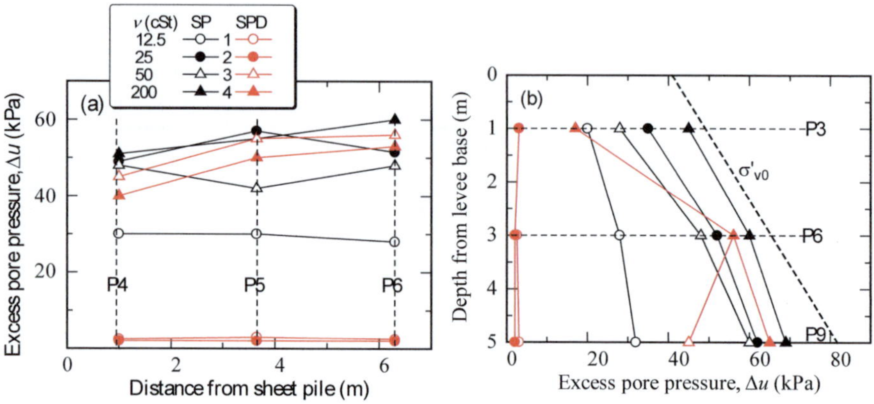

Fig. 9 Excess pore pressure distributions at the end of shaking; **a** horizontal distributions on the mid-depth of sand bed (P4, P5, and P6), and **b** vertical distributions on levee centerline (P3, P6, and P9)

all models except for those in drained condition (SPD-1 and SPD-2) in which Δu was quite low. Except for these two models, average upward gradient between P3 and P9 is approximately 0.5, which is two times higher than horizontal gradient. Allowing for the fact that the sand has horizontal permeability coefficient eight times larger than that vertical, it is reasonable to consider that amount of water flow towards the drainage member horizontally is four times larger than that upwards. This may explain that the permeability for the partially drained range is approximately four times smaller for SPD as compared to SP tests.

3 Numerical Simulation of Excess Pore Pressures

The procedure originally proposed by Seed and Booker [6] has been widely used to design vertical drains as a liquefaction countermeasure. They demonstrated that the excess pore pressure in the sand around a drain can be analyzed using the diffusion equation incorporated with generation of excess pore pressures owing to earthquake shaking, as indicated in Eq. (1). In this chapter, responses in Δu in the sand bed are simulated following the procedure with improvement made on the generation term and sand material properties used (Fig. 10).

3.1 *Excess Pore Pressure Generation*

The generation term in Eq. (1), $\partial \Delta u_g / \partial t$, is often expressed by an arcsine function for the approximation, which was derived based on undrained cyclic triaxial tests with

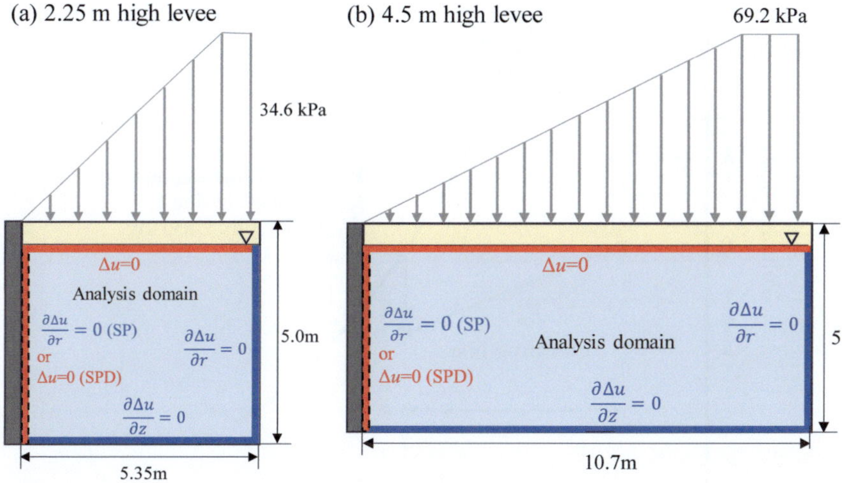

Fig. 10 Numerical analysis domains and boundary conditions. **a** 2.25 m high levee model, **b** 4.5 m high levee model

constant shear stress amplitude [6]. In this study, however, in order to simulate more closely the centrifuge models shaken with irregular base motions, the pore pressure model [10] is introduced.

The basic mechanics of densification and pore pressure generation of sand in undrained cyclic shearing were conceptually demonstrated by Martin et al. [13] as shown in Fig. 11. Volumetric strain of sand consists of recoverable strain stored by elastic deformation at grain contacts (ε_{ve}) and irrecoverable strain due to slippage at the contacts (ε_{vp}). Slippage at soil grain contacts occurs when subjected to cyclic shearing, which is the cause of plastic volumetric strain, ε_{vp},—indicated as the distance between points A and B in Fig. 11. When the undrained condition is imposed, ε_{vp} has to be compensated by the same amount of negative volumetric strain as,

$$\varepsilon_{ve} + \varepsilon_{vp} = 0 \tag{2}$$

In other words, the soil grain structure rebounds to the extent required to keep the volume constant—as indicated from points B to C on the unloading line. With the bulk modulus of water, K_w, being sufficiently high to ignore the volume change of water, generated pore pressure can be expressed using the elastic (rebound) modulus of sand, E_r, as

$$\Delta u_g = -\Delta\sigma' = -\varepsilon_{ve} E_r \tag{3}$$

Okamura [10] conducted drained cyclic shear tests and expressed ε_{vp} as,

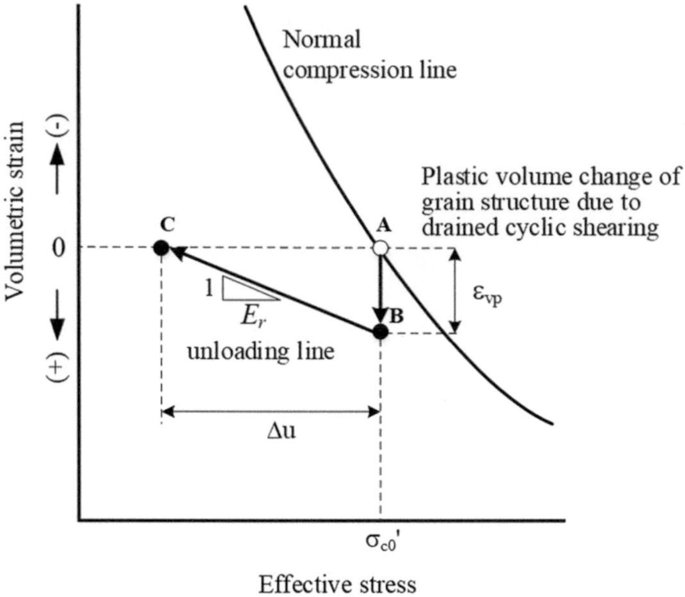

Fig. 11 Contraction by cyclic shearing converted to excess pore pressure generation

$$\varepsilon_{\mathrm{vp}} = c \times \mathrm{CSR}^{d} \times N^{b} \tag{4}$$

where CSR $= \tau_d/\sigma'_{c0}$ is cyclic stress ratio, N is number of cycles, b, c and d are material constants. Here, τ_d and σ'_{c0} are the cyclic stress amplitude and the initial effective confining pressure, respectively. A set of three parameters can be used to completely define the volume change behavior under drained cyclic shearing of a constant CSR: $(b, c, d) = (0.31, 0.051, 2.5)$ for the medium dense Toyoura sand.

In the analysis, volumetric strain in each half cycle is estimated using the time history of observed input acceleration to simulate the excess pore pressure generation of sand in the models.

The elastic (rebound) modulus of sand, E_r, was observed in triaxial tests. Specimens were isotopically consolidated at $\sigma'_{c0} = 100$ kPa, followed by unloading to 20 kPa and reloading back to 100 kPa, before and after the drained cyclic shearing. A typical response is shown in Fig. 12. Although the rebound curves were non-linear, a linear approximation was adopted for simplicity in this study. The curve was reasonably approximated by lines with $E_r = 24.0$ MPa and 13.5 MPa for the ranges from 100 to 50 kPa and 50 kPa to 25 kPa, respectively.

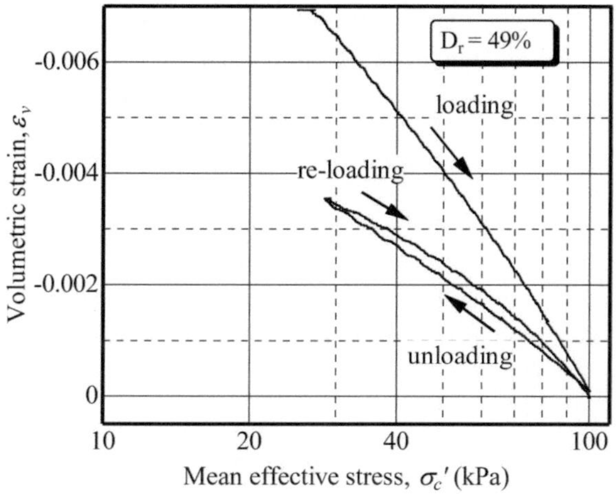

Fig. 12 Isotropic loading–unloading-reloading curves of medium dense Toyoura sand

3.2 Coefficient of Volumetric Compressibility

Minaka et al. [9] examined the volumetric strain that occurred during re-consolidation after excess pore pressure ratios, $\Delta u/\sigma'_{c0}$, were generated in triaxial testing, where σ'_{c0} denotes the initial effective confining stress. They confirmed that the re-consolidation volumetric strain (ε_v) increased linearly as $\Delta u/\sigma'_{c0}$ increases, with ε_v being practically the same irrespective of the initial effective stress σ'_{c0} for the range of $\Delta u/\sigma'_{c0} < =$ 0.75. Lee and Albaisa [7] reported similar test results for several types of sand in which the volumetric strain was almost the same irrespective of the effective confining pressure for the range of $\Delta u/\sigma'_{c0}$ lower than 0.6. They also found that the CSR in cyclic shearing to generate a certain $\Delta u/\sigma'_{c0}$ did not affect the reconsolidation volumetric strain.

Assuming the linear relationship of the swelling-reconsolidation line with slope C_s on the e–log p' plane, where e indicates the void ratio and σ'_m represents the mean effective stress, m_v can be expressed as,

$$m_v = -\frac{C_s}{(1+e)\sigma'_m} \tag{5}$$

In fact, as indicated in Fig. 12, it is reasonable to assume that the slope C_s of the medium dense Toyoura sand due to isotropic unloading is constant for the first approximation. For one-dimensional condition of sand in level surface, volumetric change of sand is not isotropic but one-dimensional.

Allowing for the anisotropic stress condition for $K_0 = 0.5$, m_v was determined as,

$$m_v = \frac{1}{E_r} = -\frac{C_s}{\sigma'_{v0}}\frac{1.5}{1+e} = -\frac{0.00652}{\sigma'_{v0}} \quad (\text{in m}^2/\text{MN}) \tag{6}$$

Equation (1) was solved in a finite difference scheme with the above-mentioned soil parameters obtained from laboratory tests. Two-dimensional analysis domain with 5.6 m wide and 4.5 m deep as shown in Fig. 10a. Coefficients of permeability of sand for vertical and horizontal directions were set independently as shown in Table 1. The initial vertical effective stress σ'_{v0} is estimated assuming the embankment height just above the soil considered. The hydraulic gradient across the boundaries is set to zero at the bottom and the centerline of the domain to represent the no-flux boundary, whereas Δu is zero at water table (drained boundary). On the surface of sheet pile, drainage boundary and no-flux boundary conditions were set for the sheet pile with and without drainage member, respectively.

3.3 Numerical Results and Comparisons with Test Observations

Figures 13 and 14 show typical results of numerical simulation compared in the form of time history of Δu with test observations. For SP-2, simulated Δu reached the initial overburden stress (σ'_{c0}) everywhere in the bed, with Δu being higher at the location of higher $\sigma'_{c0,}$ whereas the horizontal distribution was found more or less uniform in the test. Except for this upper bound Δu, the analysis simulated observed Δu satisfactory well. For SPD-2, the analysis simulated observed Δu quite well throughout the shaking. Not only vertical but horizontal hydraulic gradients are apparent which allowed volume of water to be expelled from sand bed and accelerated Δu to dissipate. After the peak was reached at t = 10 s, Δu dissipated swiftly and tended to be somewhat lower than those observed. This is largely due to the fact that the pore pressure cells in the model subsided with surrounding sand, which was evident from locations of the cells observed after the tests.

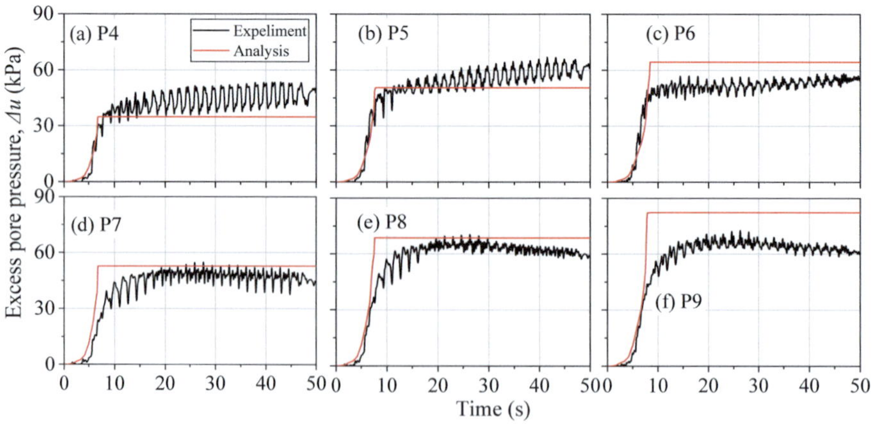

Fig. 13 Comparisons between observed and analyzed Δu for SP-2

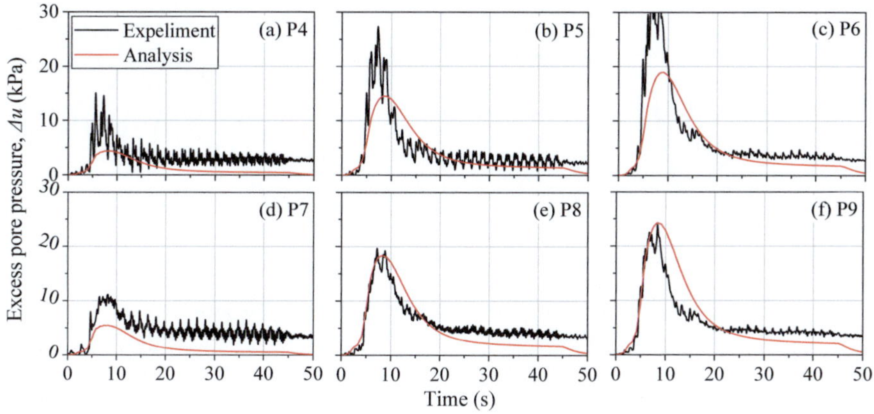

Fig. 14 Comparisons between observed and analyzed Δu for SPD-2

Simulated changes in Δu time history at P9 with the sand permeability are presented in Fig. 15. Δu of all the models without drainage member (SP-1, 2, 3, and 4) and less permeable sand models with drainage member (SPD-3 and SPD-4) are more or less similar; Δu increases swiftly at the beginning of shaking and levels off after several seconds. Further increase in the permeability significantly accelerates dissipation of Δu both in the centrifuge models and numerical simulations. Figure 16 shows evolution of excess pore pressure ratio with sand permeability. It may be noticed that sand permeabilities where Δu begin to decrease with increasing permeability tend to be higher in the simulation; Δu decreased at $k_z = 8 \times 10^{-4}$ (m/s) and $k_z = 3.2 \times 10^{-3}$ (m/s) for test without drainage ability (SP) and numerical simulation, respectively. This is also the case for the sheet pile with drainage ability that Δu decreased at $k_z = 2 \times 10^{-4}$ (m/s) and $k_z = 4 \times 10^{-4}$ (m/s) for test and simulation, respectively. A possible reason to cause the difference is the upper bound value of Δu assumed in the analysis. The higher upper bound values set in the analysis result in larger ε_{ve} and consequently longer time needs to allow larger amount of water to drain. Although the differences, the overall trend of change in simulated r_{u_max} compares well with those observed in the tests irrespective of the drainage ability of the sheet pile.

3.4 Effect of Levee Size

In this study, seismic responses of the levee of 2.25 m high were modeled and effects of the drainage ability equipped to the sheet pile and the sand permeability are investigated through the centrifuge tests and the numerical analysis. Being typical in urban areas near river mouths, levees in Japan are mostly higher than that modeled

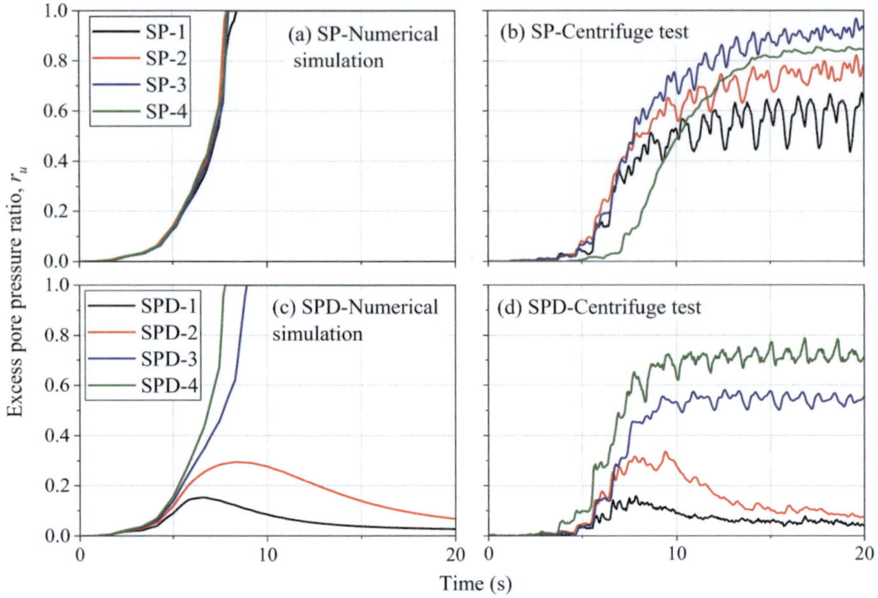

Fig. 15 Change in Δu responses with the permeability of sand at the bottom centerline (P9)

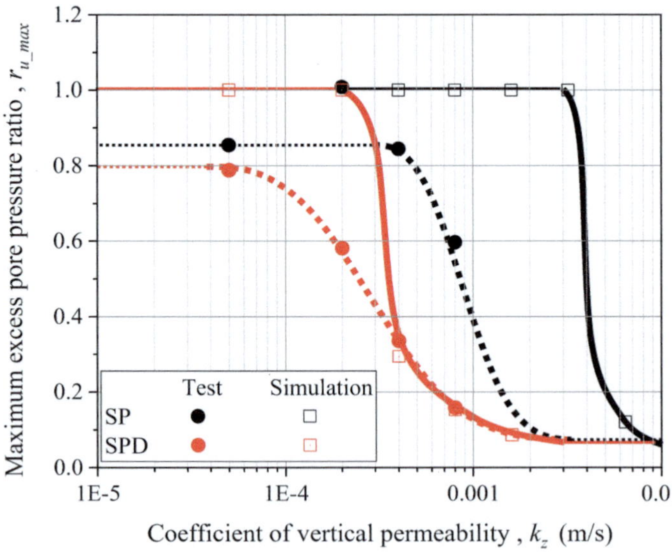

Fig. 16 Relationship between representative excess pore pressure ratio r_{u_max} at the bottom centerline (P9) and coefficient of permeability

in the former sections. It is therefore of interest to analyze the response of a larger levee.

In this section, the pore pressure responses of the medium dense 5 m deep Toyoura sand deposit with an overlying levee two times large both in height and width is analyzed (see Fig. 10b). Except for width and vertical stresses subjected to the sand surface, all conditions are identical to that described in the previous sections, including sand material properties, drainage boundary conditions and the input base acceleration.

Contours of $r_{u\_\max}$ in the sand beds both sheet pile with drainage ability for the two levees are shown in Fig. 17. Figure 18 depicts evolution of excess pore pressure ratio with sand permeability. The size of levee in this analysis affects the excess pore pressure behavior in two ways. On the one hand, the higher the effective overburden pressure at the surface of sand bed and initial effective stress in the bed are, the lower the cyclic stress ratio in the bed resulting in smaller $\varepsilon_{\mathrm{ve}}$ and Δu_{g}. This effect can be clearly seen in Fig. 18 that, for models without drainage ability (SP), $r_{u\_\max}$ for the higher levee is lower and the curve moves leftward. On the other hand, the length of horizontal drainage path is longer for the larger levee for sheet pile with drainage ability (SPD), which degrades the effect of drainage ability on the sheet pile. Difference in Δu owing to the sheet pile drainage ability is apparent in a narrower range of $4 \times 10^{-4} < k_z < 4 \times 10^{-3}$ (ten times difference in k_z) for the larger levee, while the difference is evident for the smaller levee in a wider range $2 \times 10^{-4} < k_z < 8 \times 10^{-3}$ (40 times difference in k_z). It can be concluded that the drainage ability equipped to the sheet piles is more effective for smaller levees for which the length of horizontal drainage path is short.

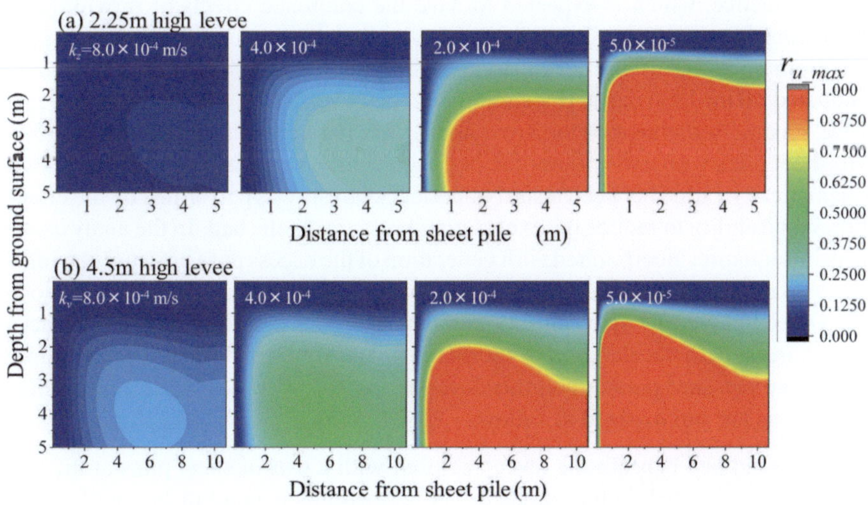

Fig. 17 Effects of levee size on contours of $r_{u\_\max}$ for sheet pile models with drainage ability

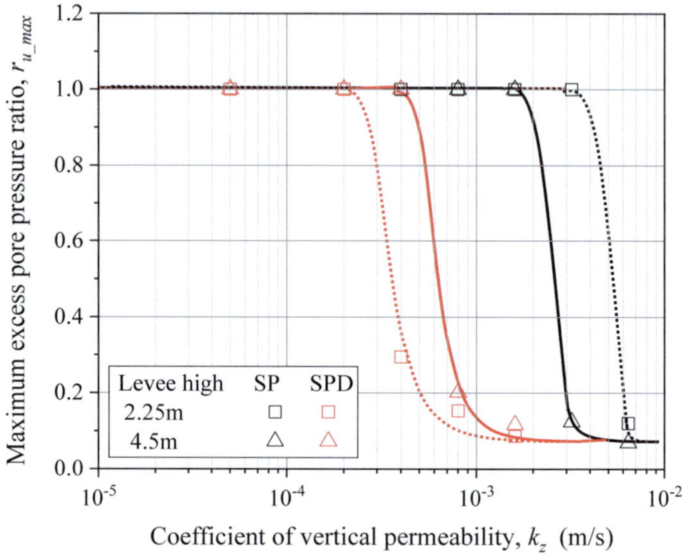

Fig. 18 Relationship obtained from numerical analysis between r_{u_max} at the bottom centerline and coefficient of permeability for levees of different size

4 Concluding Remarks

Sheet piles with additional member with drainage ability have been developed and used in practice, which is expected to have the combined effects of providing the lateral confinement by the sheet pile and facilitating dissipation of excess pore pressure from soils below the levee. In this study, a series of centrifuge tests were conducted on medium dense foundation sand bed overlain by a 2.25 m high levee to assess seismic settlement of the levee and excess pore pressure in the bed enclosed with sheet piles with and without the drainage ability. Permeability of the sand was systematically varied between tests. Numerical analysis was performed on the effects of drainage ability to reduce the excess pore pressures in the bed. In the analysis, the diffusion equation incorporated with generation of the excess pore pressures owing to earthquake shaking was used [6]. For the generation term, the pore pressure model [10] was employed to more closely simulate the pore pressure generation due to random input base acceleration. Experimental data were used to verify the validity of the pore pressure prediction procedure for soil below levees. Main findings and conclusions are summarized as follows.

- In the sand bed beneath the levee enclosed with a pair of sheet piles at the both toe, horizontal distribution of excess pore pressures is more or less uniform for cases without drainage ability. While for the case of sheet pile with drainage member, hydraulic gradient towards the sheet pile was evident which accelerated dissipation of excess pore pressure in the bed.

- Effects of drainage member is manifest when the coefficient of permeability is higher than a certain value. This threshold permeability was approximately ten times lower for models with the drainage member than that without the drainage member.
- For the range of the coefficient of permeability higher than the threshold value, crest settlement was large and almost constant irrespective of existence of the drainage member. For the range higher than the threshold value, crest settlement reduced significantly with increasing the permeability. The crest settlement decrease with the permeability is more rapid than the excess pore pressure, because sand does not lose its strength and stiffness significantly until r_u becomes higher than a certain level. The crest settlement changes more sensitively to the soil permeability than r_{u}.
- Numerical simulations were performed on the excess pore pressures of the tests using the procedure originally proposed by Seed and Booker [6]. The pore pressure generation model [10] was used to simulate the irregular input acceleration time history. It was confirmed that the numerical analysis simulated observed Δu for all the models reasonably well.
- The size of levee has effects on the excess pore pressure behavior in sand bed in two ways. One is the initial effective stress in the bed (σ'_{v0}). As increasing σ'_{v0}, the cyclic stress ratio and thus, $r_{u\_\max}$ decreases, and effects of drainage appear at the lower permeability. The other is the length of horizontal drainage path being longer for the larger levee to the sheet pile with drainage ability, which degrades the effect of drainage member on the sheet pile. Consequently, the drainage ability on the sheet pile becomes less effective as the levee size increase. On the contrary, increase in levee size decreases $r_{u\_\max}$ in a certain range of sand permeability.
- The drainage ability equipped to the sheet piles is more effective for smaller levees for which the length of horizontal drainage path is short.

References

1. Koga Y, Matsuo O (1990) Shaking table tests of embankments resting on liquefiable sandy ground. Soils Found 30(4):162–174
2. Sasaki Y, Towhata I, Tokida K, Yamada K, Matsumoto H, Tamari Y, Saya A (1992) Mechanism of permanent displacement of ground caused by seismic liquefaction. Soils Found 32(3):79–96
3. Japanese Geotechnical Society (1998) Remedial measures against soil liquefaction. Balkema, A.A
4. Adalier K, Elgamal A-W, Martin GR (1998) Foundation liquefaction countermeasures for earth embankment. J Geotech Geoenviron Eng 124(6):500–517
5. Okamura M, Matsuo O (2002) Effects of remedial measures for mitigating embankment settlement due to foundation liquefaction. Int J Phys Modell Geotech 2(2):1–12
6. Seed H, Booker J (1977) Stabilization of potentially liquefiable sand deposits using gravel drains. J Geotech Eng Div 103(GT7):757–768
7. Lee K, Albaisa A (1974) Earthquake induced settlements in saturated sands. J Geotech Eng Div 100(GT4):387–406

8. Japanese Geotechnical Society (2004) Remedial measures against soil liquefaction. Maruzen, Tokyo Japan (in Japanese)
9. Minaka U, Okamura M, Ono K (2021) Verification of effectiveness and design procedure of gravel drains for liquefaction remediation. Soils Found 61(5):1191–1206
10. Okamura M (2022) Insight into excess pore pressure generation leading to liquefaction of sand with stress history under saturated and unsaturated conditions. Soils Found 62(4):101187
11. Okamura M, Inoue T (2012) Preparation of fully saturated models for liquefaction study. Int J Phys Modell Geotech 12(1):39–46
12. Tan TS, Scott RF (1985) Centrifuge scaling considerations for fluid particle systems. Geotechnique 35(4):461–470
13. Martin GR, Finn WDL, Seed HB (1975) Fundamentals of liquefaction under cyclic loading. J Geotech Eng Div 101(5):423–438

Evaluation of Wide Ground Deformation by InSAR Time Series Analysis

Yasuhiro Mitani, Ruolin Pan, Hiroyuki Honda, and Yuki Kubotsu

Abstract Ground deformation has been measured by leveling and GNSS (Global Navigation Satellite System). However, these methods are limited in the number of measurement points, and they may overlook deformation in the area in which no leveling data is available. InSAR (Interferometric Synthetic Aperture Radar) time series analysis can measure ground deformation using SAR (Synthetic Aperture Radar) satellite data, regardless of day and night or weather conditions. Case studies have been reported using this method for analyzing land deformation. However, the applicability of this method to ground deformation monitoring, including accuracy verification, has yet to be sufficiently studied. In addition, satellite observation results cannot be directly compared with leveling survey results because they distort the satellite's line of sight. In this study, we considered the method of deformation observation by performing InSAR time series analysis and selected SBAS (Small Baseline Subset)-InSAR analysis as one of the time series analysis methods to evaluate the large area of ground deformation. The vertical displacement was obtained using 2.5-D analysis, and its accuracy was verified with leveling survey results to study the applicability of SAR satellite in large-area comprehensive land deformation monitoring.

Keywords InSAR · Land deformation · SBAS-InSAR · Sentinel-1

Y. Mitani (✉) · H. Honda
Disaster Risk Reduction Research Center, Faculty of Engineering, Kyushu University, Fukuoka, Japan
e-mail: mitani@doc.kyushu-u.ac.jp

R. Pan
Graduate School of Engineering, Kyushu University, Fukuoka, Japan

Y. Kubotsu
Faculty of Engineering, Kyushu University, Fukuoka, Japan

H. Hazarika et al. (eds.), *Multidisciplinary Approach for Disaster Resilience and Sustainable Technologies*, Lecture Notes in Civil Engineering 642,
https://doi.org/10.1007/978-981-96-8618-6_13

1 Introduction

Leveling and GNSS measurement methods are used to monitor the ground deformation caused by pumping up the underground water, according to the production of gas. However, due to cost and workforce constraints, the number of measurement points is limited, and it is possible to miss local ground deformation. Therefore, a method called InSAR time series analysis using SAR satellites has begun to be used, which enables monitoring over a wide area and long term. In InSAR analysis, the amount of variation is calculated from two images taken at different times using the phase difference between the radio waves of the two periods. However, this method includes errors due to atmospheric effects and satellite orbit deviation. InSAR time series analysis can reduce the errors in InSAR analysis [1]. However, deformations in the vertical and horizontal directions of the ground cannot be observed in the deformations of satellite LOS (line of sight) obtained through analysis [2]. In this study, we use InSAR time series analysis method and 2.5-D analysis method to evaluate the vertical deformation of a wide ground area [3]. The variation tendency of ground deformation is compared by combining it with the leveling data [4].

2 Study Area and Data

Figure 1a shows this study's Sentinel-1 satellite image set and its observation direction. Figure 1b shows the study area of this study, as well as the location of leveling points and natural gas production wells in the study area. The target area is along the coast of Miyazaki Prefecture, extending approximately 20 km from north to south and 10 km from east to west. In Miyazaki city, along the coast of Miyazaki Prefecture, natural gas exploration and development have been conducted since shortly after World War II. In the Sadohara area, which is the study area of this study, natural gas development began in 1975, and production continues to this day [5].

The natural gas exploitation method in this area is to pump up underground water. For the development safety of this area, it is necessary to observe the ground deformation. In this area, the deformation has been predicted based on level surveys to correctly predict the relationship between development and deformation and promote safe development. With the expansion of the natural gas development area, it is essential to observe ground deformation in the area for future development properly.

In this study, we use the Sentinel-1 satellite, which employs C-band (wavelength of about 5.5 cm) radar. This satellite is an earth observation satellite launched by the European Space Agency (ESA) and operated by two satellites, Sentinel-1A (S1A) and Sentinel-1B (S1B). These satellites orbit at an altitude of 693 km with a period of 98.74 min, and both S1A and S1B have been observing at intervals of about 12 days within their observation range. For this study, we use data from approximately five years, from January 2017 to December 2021 (S1A: 63 data, S1B: 61 data), for which Sentinel-1A and 1B both have data. The properties of Sentinel-1 data are shown

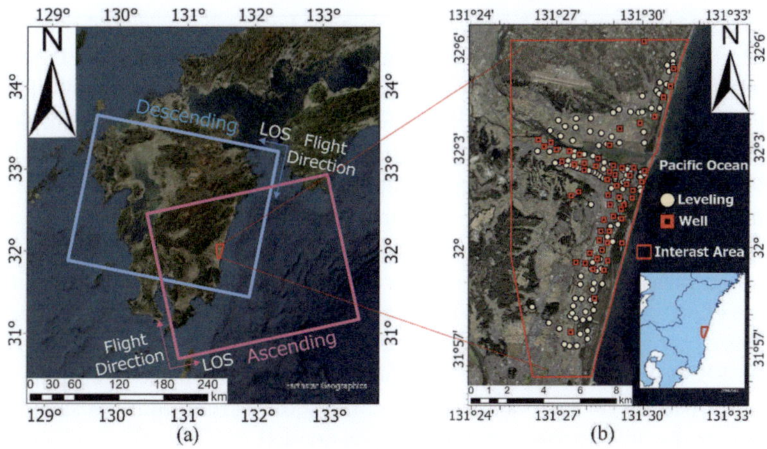

Fig. 1 **a** Sentinel-1 satellite images set used and its observation direction; **b** location of the study site (target area)

Table 1 Properties of Sentinel 1A and Sentinel 1B data

Satellite	Sentinel 1A	Sentinel 1B
Band	C	C
Orbit	Descending	Ascending
Date From To	1 January 2017 26 December 2021	3 January 2017 8 December 2021
Master image	24 March 2019	31 December 2019
Number of scenes	63	61
Incidence angle	32.73°–32.74°	31.88°–32.52°
Spatial resolution	5 m × 20 m	5 m × 20 m
Revisit cycle	12 days	12 days

in Table 1. Even if the data are within the observation period, the data with poor interference after interference processing, data with deviations in the observation coverage, and data with turbulence in the satellite images are excluded. In this study, the analysis is performed using the data set of VV polarization of IWS mode, which is the most common mode for Sentinel-1 land-based observations.

3 Methodology

SAR satellite emits a microwave by itself to observe the reflected electric wave from the object and measures the properties of object's surface, size, and shape according to the intensity of the reflected electric wave. In addition, the approximate distance to the object can be determined by measuring the time before the irradiated electric wave is reflected and returned. The distance data can be obtained by analyzing the "phase" of the received radio wave. InSAR is a method to determine the amount of ground deformation by reading the phase difference of two radio waves from multiple satellite data observed at almost the exact location in different periods [6]. However, since InSAR measures the distance from the satellite's line of sight directly to the object from the phase difference obtained from multiple observations, it can be affected by orbital data, water vapor, ionosphere, and DEM (Digital Elevation Model) [7]. To reduce their influence, an InSAR time series analysis has been studied to eliminate the cause of error in jamming SAR using a large amount of accumulated observation data. This experiment uses the SBAS-InSAR analysis method in InSAR time series analysis. SBAS-InSAR analysis is a method to generate multiple interference pairs with short observation intervals and vertical baselines from multiple satellite data and estimate the deformations in each observation [8].

In this study, the SARScape module of ENVI 5.6 software provided by Harris Geospatial Solutions is used for SBAS-InSAR analysis. The seven steps of the SBAS-InSAR analysis workflow are as follows [9]:

1. Creating a connection diagram defines interference pairs to perform interferometric processing on multiple images.
2. Generating differential interferograms: spectral migration and adaptive filtering.
3. Phase unwrapping: using conventional 2D methods or combined 3D methods.
4. Orbital thinning and refocusing.
5. First estimation of mean displacement.
6. Atmospheric phase shielding.
7. Final estimation of mean displacement.

In this study, we achieve a ground resolution of 15 m using a multi-view factor with a multi-looking factor of four in the range and one in the azimuth. The 10 m DEM (Digital Elevation Model) removes the terrain phase.

The ground deformation obtained from InSAR is the deformation of the distance from the satellite to the surface, that is, the one-dimensional component of the direction of the satellite LOS, due to the limitation of SAR observation. However, the leveling data cannot be compared with the deformation in the direction of the satellite line of sight due to the observed vertical deformation of the ground (uplift and down). Therefore, we carried out a 2.5-D analysis. By combining the observation results of S1A and S1B in two different directions, this method separated the quasi-up and down components and quasi-east–west components in the plane and regarded the quasi-up and down components as deformations in the vertical direction, which can be compared with the horizontal measurement in the vertical direction.

4 Results and Discussions

4.1 Results of SBAS-InSAR Analysis

SBAS-InSAR analysis was carried out for S1A and S1B data, and 260,107 analysis points were obtained for S1A and 258,419 for S1B. The 2.5-D analysis is conducted for the two results to estimate each point's vertical ground deformation rate. 231,380 analysis points were obtained. Figure 2 shows the annual vertical deformation rate of 2.5-D results obtained by plotting the color-coded vertical deformation rate of each analysis point. The cold color indicates the area of land deformation, while the warm color indicates the area of land uplift.

The analysis points processed by SBAS-InSAR and 2.5-D analysis are widely distributed, and the deformation of most areas can be detected. However, from the location distribution of the analysis points, the areas with many artificial objects (buildings) are densely covered by the analytic points. However, there are still some areas where deformations are only partially observed. That may be because the observation band of the satellite used is C-band, and its wavelength is short, so it is difficult to observe the deformations in low-scattering and unstable signal areas caused by vegetation deformations such as forests, fields, and mountains. The vertical variation obtained from the 2.5-D analysis shows that the settlement distribution area near the estuary of Hitotsuse River is vast. From the analytical results, the area is gradually sinking at a rate of – 9.89 mm/year to – 0.54 mm/year on average every year. Some uplifts can be seen in the north and southeast of the study area. From the analysis results, the maximum uplift speed can reach 9.4 mm/year.

4.2 Verification of Accuracy by Leveling

To verify the accuracy of SBAS-InSAR analysis, it is compared with the horizontal measurement results (131 points). During the comparison, the analysis point is not necessarily at the exact location of the leveling point, so it is compared with the nearest analysis point within 40 m of each leveling point. As a result, the average absolute value of the annual variation rate error with the horizontal measurement result is 1.18 mm/year. Here, the comparison results of the annual variation rate of the analysis results and the leveling results are shown in Fig. 3. The dotted line in the figure is the line that the analysis result is consistent with the leveling result, and the straight line is the line with an error of ± 1 mm/year. The analysis results tend to show values slightly larger than the horizontal measurement results, with 66 points (50.4%) within ± 1 mm/year and 5 points above ± 3 mm/year.

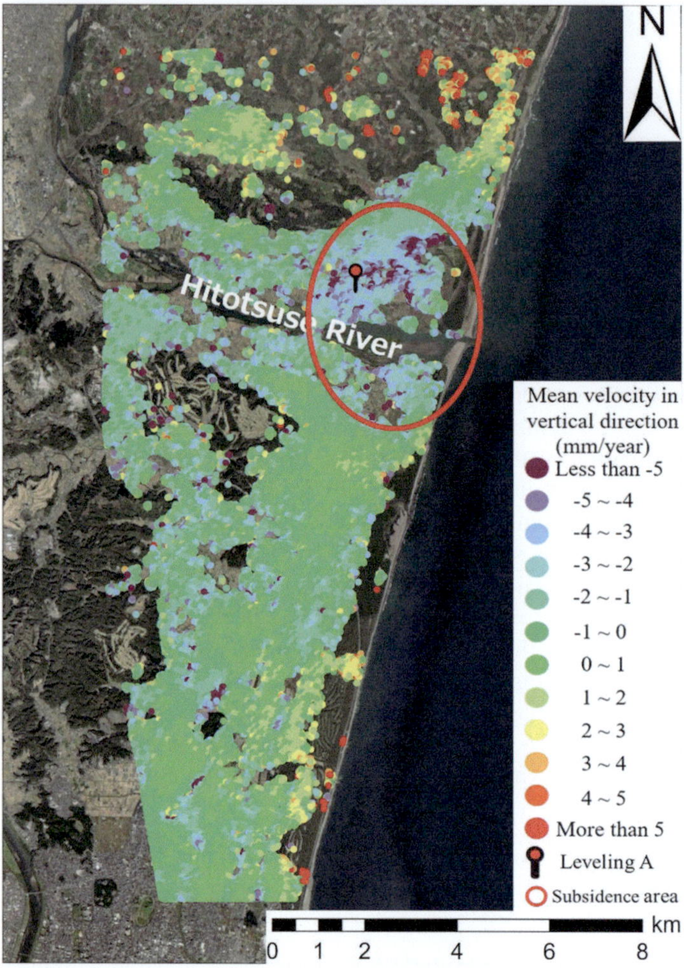

Fig. 2 Deformation velocity in the vertical direction from January 2017 to December 2021

4.3 Compare Leveling and Analytical Results on Time Series Analysis

In addition, to compare the changes in time series, Fig. 4 shows the changes of horizontal measurement points and analysis points at point A in Fig. 2. Both results calculate the cumulative change based on January 1, 2017. The analysis results slowly subside within five years and repeat some differential deformation and uplift. In addition, the leveling results also tend to sink, and the accumulated variation also shows approximately the same value. The annual variation rate analysis result is −

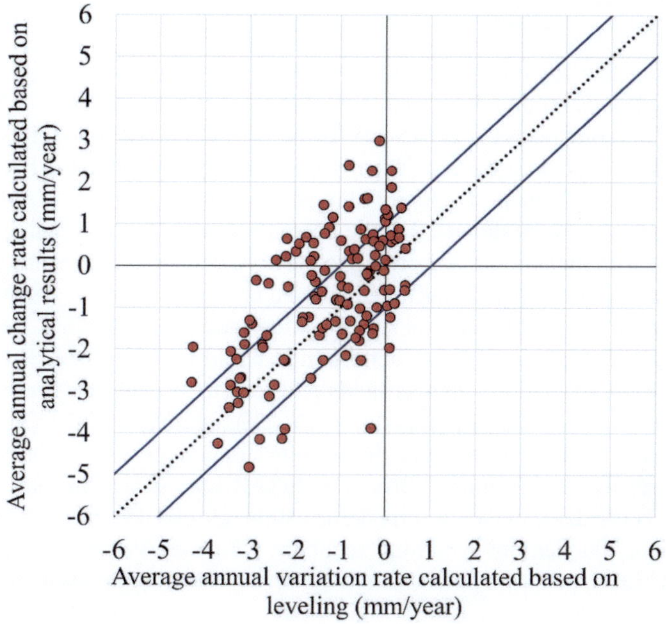

Fig. 3 Scatter plots of leveling and analysis results

3.03 mm/year, and the leveling result is − 3.28 mm/year, showing approximately the same value.

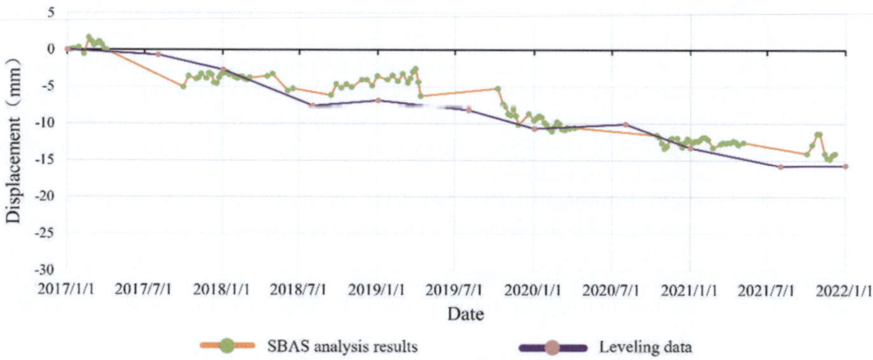

Fig. 4 Variation of satellite observation results and field survey results, in mm

5 Conclusions

This study aims to study the applicability of SBAS-InSAR analysis using Sentinel-1 satellite in wide area ground deformation monitoring in the coastal area of Miyazaki Prefecture. This study uses the SBAS-InSAR method in InSAR time series analysis to measure the ground deformation. Then, the 2.5D analysis method has been used to obtain the ground deformation velocity of most areas in the study area from 2017 to 2021. The analysis results showed that the fluctuation could be observed outside the horizontal measurement point. According to the horizontal measurement results, the ground deformation could be determined with an error accuracy of 1.18 mm/year.

Regarding the observable range of surface deformation in urban areas, the analysis point density was high, and planar deformation could be confirmed. However, in areas covered by vegetation, such as fields and forests, it could be confirmed that there were fewer deformation points due to the characteristics of the C-band. The time series change results showed that the satellite analysis results are consistent with the leveling results, with minimal error. The InSAR time series analysis results of satellite data could detect the long-term deformation rate and determine the trend and accumulation of fluctuations with high accuracy, which showed that SBAS-InSAR analysis had high application potential in large-area ground deformation monitoring.

References

1. Ishitsuka K, Matsuoka T (2016) Accuracy evaluation of persistent Scatterer interferometry using ALOS/PALSAR data: a case study of surface displacement in the Kujukuri Plain, Chiba Prefecture. J Remote Sens Soc Jpn 36:328–337 (2016). https://doi.org/10.11440/rssj.36.328
2. Morishita Y (2016) Comparison of PSI results between L- and C-band SAR data. J Geod Soc Jpn 61:157–166 (2016). https://doi.org/10.11366/sokuchi.61.157
3. Asaka T, Nonaka T, Hashiba H, Iwashita K, Sugimura T (2018) ALOS/PALSAR SBAS analysis of surface deformation in the Pacific side of Chiba Prefecture, Japan. Trans Jpn Soc Aeronaut Space Sci Aerosp Technol Jpn 16:593–598 (2018). https://doi.org/10.2322/tastj.16.593
4. Xu H (2012) Application of GPS-RTK technology in the land change survey. Procedia Eng 29:3454–3459. https://doi.org/10.1016/j.proeng.2012.01.511
5. Susumu K, Amane W, Hirotsugu I (2011) Geochemistry of natural gas and formation water from water-dissolved gas fields in Miyazaki Prefecture. J Jpn Assoc Pet Technol 76:244–253. https://doi.org/10.3720/japt.76.244
6. Yamanaka M, Morishita Y, Osaka Y (2013) Detection of ground subsidence by InSAR time series analysis. GSI Jiho J Geospatial Inf Auth Jpn 1–14. https://www.gsi.go.jp/common/000079938.pdf
7. Interferometric SAR Error | GSI. https://www.gsi.go.jp/uchusokuchi/sar_error.html. Last accessed 09 Nov 2022
8. Tizzani P, Berardino P, Casu F, Euillades P, Manzo M, Ricciardi GP, Zeni G, Lanari R (2007) Surface deformation of Long Valley caldera and Mono Basin, California, investigated with the SBAS-InSAR approach. Remote Sens Environ 108:277–289. https://doi.org/10.1016/j.rse.2006.11.015
9. Closson D, Holecz F, Pasquali P, Milisavljevic N (2014) Land applications of radar remote sensing. BoD—Books on Demand

Assessing the Hydro-chemo-mechanical Behavior of Marble Dust-Bentonite Mixes to Be Used as a Geomaterial

Ankush Kumar Jain, Arvind Kumar Jha, and Mohammad Parwez Akhtar

Abstract The utilization of a huge amount of waste generated from mining industries is vital for the sustainable and eco-friendly design of construction projects such as roads, embankments, and buildings. The World's largest marble production, i.e., Rajasthan state of India is producing approximately 95% of the total marble in the country. The unscientific disposal of about 70% of marble waste (i.e., 15–20 lac tons) generated from 4000 marble mines in Rajasthan poses a major environmental threat due to its suspension in the air, increase in alkalinity of soils, water logging due to the reduction in porosity of the topsoil, etc. Marble dust (MD) has several advantageous properties such as coarser-sized particles, high shear strength (friction angle up to 44°), and non-plastic behavior which are alike with sand. Further, depletion of natural sand has forced to find suitable alternative material for its application as a geomaterial. Despite its unique characteristics of marble dust, the proper attempt has not been made to investigate the suitability of marble dust to be used alone or amendment with other additives such as bentonite (B), chemical stabilizers, etc. for geotechnical projects. The preliminary objective of present work is to find out its potential to be used in subgrade soil preparation. Detailed geotechnical investigations such as physical and engineering properties, physicochemical examination (pH and electrical conductivity) and microanalyses (mineralogical and microstructural) have been investigated in order to determine optimize the MD content for soil improvement. The secondary objective of the present work has been aimed to examine the effectiveness of marble dust amended with bentonite to be used as a landfill liner material in the substitution of sand–bentonite (S–B) mix. The geotechnical properties (i.e., swelling percentage and permeability) of MD–B mixes that are required to satisfy landfill liner design and construction are compared with S–B

A. K. Jain
Poornima University, Jaipur, India

A. K. Jha (✉)
Indian Institute of Technology Patna, Patna, India
e-mail: jhaarvind@iitp.ac.in

M. P. Akhtar
Manipal University Jaipur, Jaipur, India

H. Hazarika et al. (eds.), *Multidisciplinary Approach for Disaster Resilience and Sustainable Technologies*, Lecture Notes in Civil Engineering 642,
https://doi.org/10.1007/978-981-96-8618-6_14

mixes. It is well known that the objective of engineered landfill liners is to control the migration/leakage of hazardous leachate solutions generated from dump wastes into the groundwater. Also, the interaction of liner material with hazardous leachates may lead to the alteration in the desired properties. Hence, the effects of electrolyte solutions (NaOH and NaCl) of various concentrations on the behavior of optimized MD–B mixes are investigated to confirm the efficacy and longevity potential to be used as an alternative to S–B mixes in landfills liner. Mechanisms of alteration in the properties of both mixes subjected to electrolyte concentration are brought out clearly through physicochemical and microanalyses.

Keywords Bentonite · Geomaterial · Marble dust · Landfill liner · Waste

1 Introduction

Geomaterials such as clay, granular materials, sand, fly ash, geo-synthesis, lime, cement dust, etc. are utilized to securely build civil engineering structures. Infrastructure growth has raised construction costs for traditional materials. Several geotechnical applications of geomaterials are mostly observed in landfill liners, roadway pavements, retaining walls, slope stability, excavation and supporting ground structures, natural and artificial fills, and so on. Thus, present geotechnical construction should use industrial byproducts as geomaterial without affecting the environment.

Rajasthan, a state of India, produces a largest amount (i.e., 95% of total amount) and is considered as World's largest marble production. The unscientific disposal of about 70% of marble waste (i.e., 15–20 lac tons) generated from 4000 marble mines in Rajasthan pollutes land and air, affecting humans and ecosystems. Literature reported that marble dust has low cohesion and high internal friction [1]. Marble dust has a higher CBR than fly ash, making it appropriate for road construction [1]. Marble dust is water-resistant and porous [2]. Marble dust has low liquid and plastic limits and thereby, poor plasticity index [2]. Tozsin et al. [3] suggest marble dust neutralizes acidic soils. According to literature, appropriate marble dust content increases unconfined compressive strength (UCS) of the red and black cotton soil (BCS) [4]. Due to such advantageous characteristics, marble dust could be explored as a potential geomaterial to improve the geotechnical behavior of problematics soil for its applications in highway, landfill constructions, and so on.

Expansive soil is considered problematic in nature due to its swell-shrink phenomenon upon temporal variations [5]. Geotechnical engineers are looking for strategies to treat and stabilize expanding soil [6]. Building highway pavement on expansive soil requires proper soil preparation for subgrade to control its volume change behavior. Stabilization enhances geotechnical behavior of expansive soil most often.

Most countries prefer landfilling for disposal of solid waste [7]. Leachate from dumping sites can contaminate groundwater. Thus, suitable landfill liner needs to be constructed to control the migration of contaminant leachate. Several landfill liners such as compacted clay liner (CCL), geosynthetic clay liner (GCL), sand–bentonite, etc. are being constructed to reduce leachate migration. However, sand–bentonite liner is ideal for landfill liners due to its permeability, strength, and plasticity. Sand is scarce and regulated, causing concern. Thus, alternative material needs to be investigated to replace the utilization of natural sand.

Hence, present research primarily explores the potential utilization of marble dust as geomaterial for two different geotechnical applications, i.e.; (i) subgrade soil for highway pavement and, (ii) developing a secure landfill liner. The first objective is achieved by performing extensive experimental works; scratching from the detailed characterization of marble dust obtained from mining sites and its interaction with expansive soil to find out its potential to be used in subgrade soil preparation. Geotechnical properties, physicochemical examination (pH and electrical conductivity) and microanalyses have been investigated in order to determine optimize the MD content for soil improvement. However, second objective is achieved by finding the geotechnical properties of MD–B mixes that are required to satisfy landfill liner design and construction which are then compared with S–B mixes. Further, effects of electrolyte solutions (NaOH and NaCl) of various concentrations on the behavior of MD–B mixes are investigated to confirm the efficacy and longevity potential to be used as an alternative to S–B mixes in landfills liner.

2 Material and Methodology

2.1 Collection of Material

The sandy soil was collected from Manipal University Jaipur, Dahmi Kalan, Jaipur district, Rajasthan, India. Black Cotton Soil (BCS) was collected from Udaipuriya village in Shivdaspura, Jaipur district, Rajasthan. Bentonite was collected from the Bikaner district, Rajasthan. The marble dust was collected from Kishangarh, Ajmer district, Rajasthan. In this study, distilled water and solution of sodium chloride (NaCl) and sodium hydroxide (NaOH) of different concentrations were used as pore fluids/electrolyte concentration. Further, the geotechnical properties of parent materials and adopted methodology are presented in Table 1.

2.2 Mineralogical, Microstructural and Chemical Composition Analysis

Mineralogical analysis of parent materials was carried out by X–ray diffraction (XRD) method (Fig. 1). The rutile, quartz, feldspar, and mica were identified as major minerals in sandy soil. Montmorillonite, aluminium oxide, and quartz were found to be most common minerals in BCS, as per XRD analysis. The existence

Table 1 Geotechnical properties of parent samples

Description	Sand	Soil	MD	Bentonite	IS Code
Sand (4.75–0.075 mm), %	98.50	11.00	96.50	8.00	IS 2720 (Part 4) [8]
Silt (0.075–0.002 mm),%	1.50	13.00	3.50	15.00	
Clay (< 0.002 mm),%	–	76.00	–	77.00	
Soil classification	SP	CH	SP	CH	
Specific gravity	2.63	2.38	2.74	2.65	IS 2720 (Part 3) [9]
Liquid limit, %	30.76	49.00	17.60	185.00	IS 2720 (Part 5) [10]
Plastic limit, %	–	29.18	–	67.33	
Shrinkage limit, %	–	11.65	–	8.05	IS 2720 (Part 6) [11]
Plasticity index, %	–	19.81	–	117.67	
MDD, gm/cc	1.58	1.62	1.87	1.12	Sridharan and Sivapullaiah [12]
OWC,%	14.44	20.83	15.03	45.76	
CBR,% (Un-soaked)	20.04	9.25	14.12	1.74	IS 2720 (Part 11) [13]
CBR,% (Soaked)	12.14	3.19	6.67	0.31	
UCS, kPa	–	238.19	–	242.83	IS 2720 (Part 10) [14]
Free swell index, %	–	70.00	–	490.00	IS 2720 (Part 3) [15]
MFSI, ml/gm	–	3.04	–	13.04	Sivapullaiah et al. [16]
pH value	8.31	8.07	8.18	7.97	IS 2720 (Part 26) [17]
EC value, ms/cm	0.13	0.15	0.11	0.28	
Cohesion (at OWC), kPa	19.00	37.51	18.00	40.50	IS 2720 (Part 13) [18]
Angle of internal friction (at OWC), degree	22.54	6.19	25.01	3.76	
Maximum swell, %	–	2.53	–	53.22	IS 2720 (Part 41) [19]
Permeability, cm/s	6×10^{-4}	1×10^{-7}	4×10^{-3}	2×10^{-12}	IS 2720 (Part 17) [20]

of dolomite, quartz, and calcite was observed as predominant minerals in marble dust. Minerals such as montmorillonite, kaolinite-montmorillonite, and quartz were detected in bentonite sample.

Microstructural examination was carried out by a Scanning Electron Microscope (SEM). Rough texture and presence of small cavities was observed on the surface of sandy soil particles (Fig. 2a). Microstructural examination of marble dust confirmed irregular and flaky shapes of particles (Fig. 2b). Microstructural examination of BCS illustrated the presence of voids having a honeycomb structure (Fig. 2c). The SEM

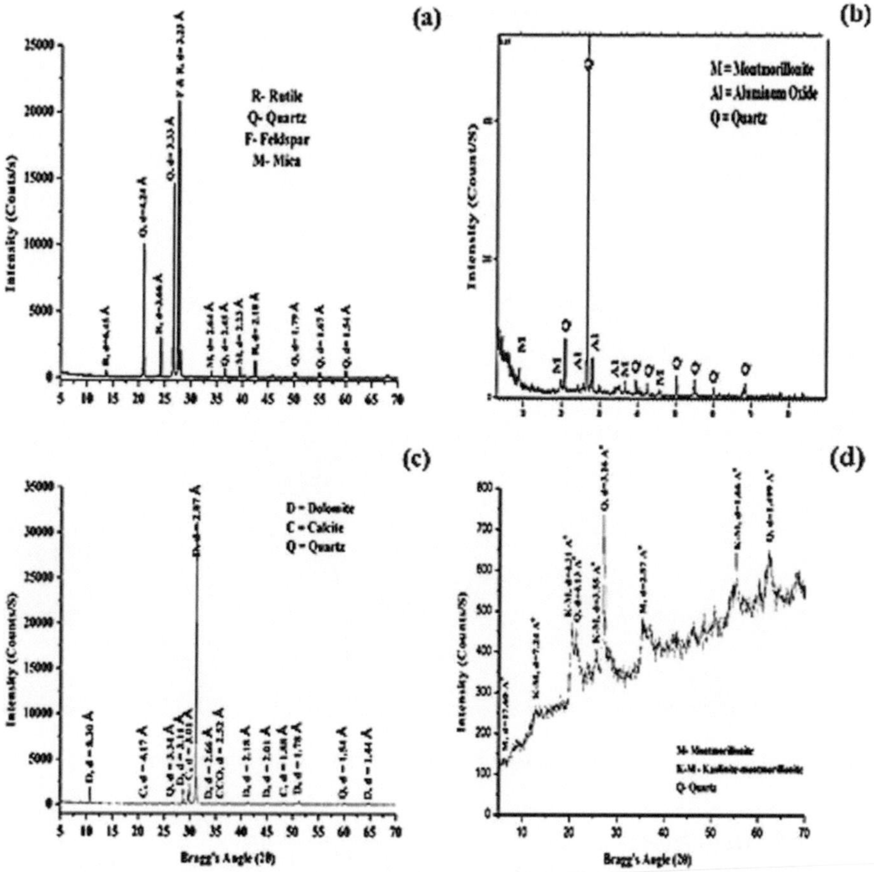

Fig. 1 XRD analyses of **a** sand **b** black cotton soil **c** marble dust and **d** bentonite [21–25]

image of bentonite sample illustrated the clay platelets arranged in a face-to-face orientation in the sample (Fig. 2d).

3 Utilization of Marble Dust for the Improvement in Black Cotton Soil for Highway Construction

This section describes the effectiveness of marble dust in enhancing the properties of expansive soil for its bulk utilization in subgrade preparation. The experimental analysis has been conducted by mixing the soil with marble dust on a dry weight basis. The mixture of soil and marble dust content is designated by using %BCS + %MD for example, 95BCS + 5MD, where 95 and 5 represent percentage of soil and marble dust, respectively.

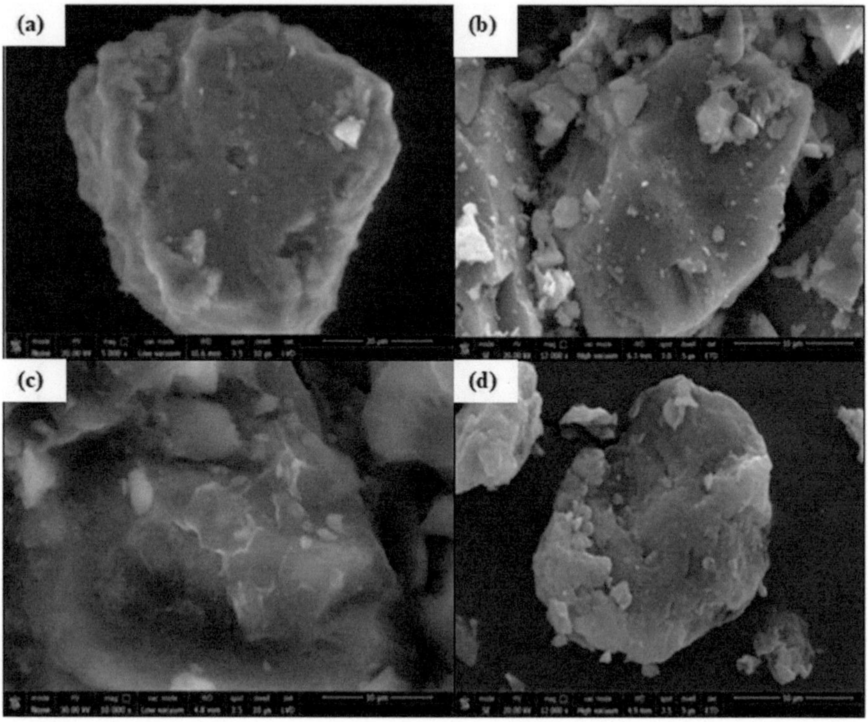

Fig. 2 SEM examinations of **a** sand; **b** marble dust; **c** black cotton soil; and **d** bentonite [21–25]

3.1 Physicochemical Analysis and Compaction Characteristics of Soil–Marble Dust Mixes

The pH of soil amended up to 15% marble dust content increases. Further increase in MD up to 40% has reduced the pH of soil. It is well known that marble dust contains calcite that enhances the salt concentration in pore fluid, resulting an alteration in pH of soil amended up to 15% marble dust [26–28]. Electrical conductivity of soil predominated with clay-sized particle reduces with an increase in amount of marble dust predominated with sand sized particles as shown in Table 2.

The compaction characteristics of soil amended up to 80% MD show an increase in maximum dry density (MDD) with reduction in optimum water content (OWC). The gradation of soil improves with addition of MD content that causes an enhancement in MDD of soil [29].

Table 2 Physicochemical analysis and compaction characteristics of soil–marble dust mixes

Mixes	pH	EC (ms/cm)	MDD (g/cc)	OWC (%)
100BCS + 0MD	8.07	0.151	1.62	20.83
95BCS + 5MD	8.53	0.149	1.643	22.79
90BCS + 10MD	8.6	0.146	1.691	19.09
85BCS + 15MD	8.65	0.143	1.691	18.83
80BCS + 20MD	8.42	0.137	1.692	17.24
75BCS + 25MD	8.2	0.132	1.704	16.07
60BCS + 40MD	7.99	0.123	2.070	9.19
40BCS + 60MD	8.14	0.119	2.110	8.96
20BCS + 80MD	8.29	0.111	2.160	8.86
0BCS + 100MD	8.18	0.106	1.870	17.6

3.2 *Unconfined Compressive Strength (UCS) and California Bearing Ratio (CBR) of Soil–Marble Dust Mixes*

It is revealed (Fig. 3) that strength of soil increases continuously with an increase in marble dust content up to 20%, irrespective of curing period. Further increase in marble dust content leads to lessen the strength of soil. However, it is interesting to note that the strength of soil enhances unceasingly with an increase in curing period at any particular marble dust content. Hence, 20% marble dust can be considered as an optimum marble dust content from strength point of view to amend the soil.

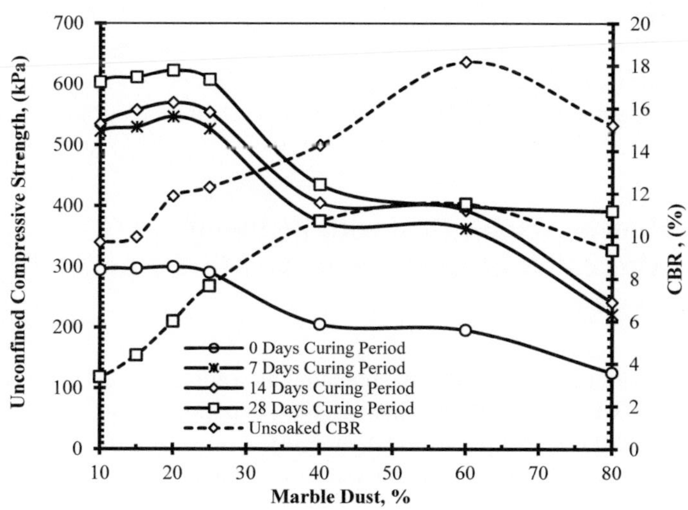

Fig. 3 UCS and CBR value of soil–marble dust mixes [21, 22]

The CBR of BCS amended with varying marble dust content up to 100% has been examined to know its effectiveness in utilizing as a subgrade material for highway construction. Further, a drastic improvement in the CBR value of BCS is observed in Fig. 3 with an addition of MD content up to 60%. The improvement in CBR value of BCS with effect marble dust content also satisfies the criteria of Indian Road Congress for the possible utilization in road construction [22, 30–32]

3.3 Analytical Analysis

The XRD analysis of soil amended with 20% MD content at curing periods of 28 days has appeared new peak at ($2\theta = 17.7°$) that indicates the formation of mineral (Fig. 4). However, the peak related to aluminium at 2θ value = 25.69° has weakened with substantial sharpening in calcite peaks at 2θ value = 31.11° and 37.51°. The alteration in mineral peaks revealed the formation of compacted soil matrix, leading to the increase in the strength with decrease in swelling percentage.

The microstructural examination of BCS + 5MD at 7 days curing period have shown the poorly aggregation of soil matrix having larger voids/pores (Fig. 5a). Further the sample subjected to 28 days curing periods has pronounced relatively denser matrix with less voids/pores (Fig. 5d), indicating healing of weak bonding between the particles with curing periods.

Figures 5b, e show the SEM images of soil with 25% MD without and with cured for 28 days which confirm the denser matrix. This is due to the improvement in bonding between soil and marble dust particles due to formation of cementing gels that result in decrease in voids/pores within sample [31]. The formation of compacted and denser matrix validates the strength improvement in sample. However, amendment of soil with higher content of MD (i.e., 80%) has shown the presence of loose particles which indicate the lack of interlocking and cohesion (Fig. 5c, f). This reduces strength of soil amended with higher MD contents at different curing periods.

4 Utilization of Marble Dust with Bentonite as an Alternative of Sand–Bentonite Mix for Landfill Liner System

This section is aimed to investigate the potential use of marble dust as a geomaterial in combination with bentonite. Further, detailed experiments on both S (Sand)–B (Bentonite) and MD (Marble Dust)–B (Bentonite) mixes with and without electrolyte concentration are conducted to examine the geotechnical behavior.

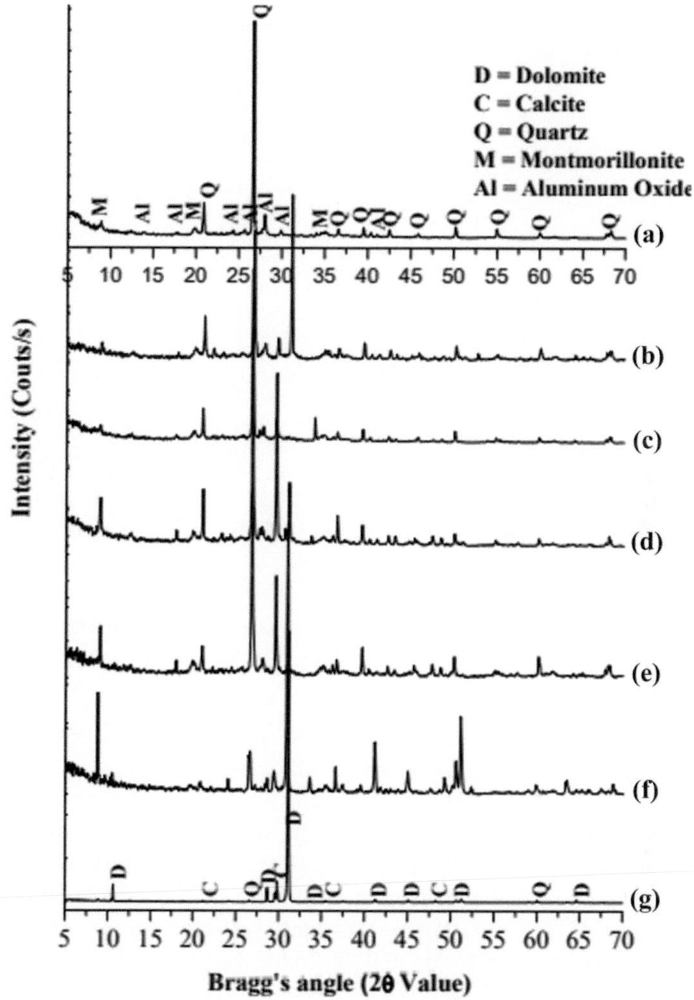

Fig. 4 XRD analysis of **a** 100BCS, **b** 95BCS + 5MD, 28 days, **c** 80BCS + 20MD, 14 days, **d** 80BCS + 20MD, 28 days, **e** 75BCS + 25MD, 28 days, **f** 20BCS + 80MD and **g** 100MD [24]

4.1 Physicochemical Analysis and Compaction Characteristics of S–B and MD–B Mixes

The pH values of sand and marble dust increase from 8.31 to 10.05 and 8.20 to 9.80 with the addition of bentonite up to 20%, respectively. Further increase in bentonite reduces the pH of the sand and marble dust. Further, the EC of sand and marble dust increases from 0.106 to 0.258 ms/cm and 0.135 to 0.273 ms/cm by addition of bentonite content up to 95%, respectively (Table 3).

Fig. 5 SEM analysis of **a** 100BCS, **b** 95BCS + 5MD, 28 days **c** 80BCS + 20MD, 14 days **d** 80BCS + 20MD, 28 days **e** 75BCS + 25MD, 28 days, **f** 20BCS + 80MD and **g** 100MD [24]

It is observed that MDD of marble dust with lower bentonite content up to 60% is observed to be greater than sand which is almost similar thereafter, indicating the higher rate of reduction in MDD.

The requirement of additional water in the mixture causes a swelling in the bentonite, and that swelled bentonite is lighter than marble dust and sand, which occupies larger space in the compaction mold, leading to the reduction in the density of the mixtures [32]. The OWC of sand and marble dust increases continuously with the addition of bentonite content up to 90% (Table 3).

4.2 Maximum Swell Percentage of S–B and MD–B Mixes Without Electrolyte Concentration

It is observed from Table 4 that the maximum swelling percentage of both sand and marble dust increases continuously with an increase in the bentonite content. The presence of Na^+ montmorillonite and the high cation exchange capacity (CEC) of bentonite yields high swelling in S–B and MD–B mixes.

Table 3 Physicochemical analysis and compaction characteristics of S–B and MD–B mixes

Mixes	pH	EC (ms/cm)	MDD (g/cc)	OWC (%)
100S + 0B	8.31	0.135	1.58	14.44
100MD + 0B	8.20	0.106	1.87	15.03
95S + 5B	8.61	0.175	1.57	15.08
95MD + 5B	9.20	0.144	1.85	15.3
90S + 10B	9.18	0.189	1.55	16.95
90MD + 10B	9.50	0.178	1.83	15.46
80S + 20B	10.05	0.22	1.54	18.54
80MD + 20B	9.80	0.183	1.78	16.03
60S + 40B	8.78	0.229	1.47	24.12
60MD + 40B	9.30	0.199	1.56	22.48
40S + 60B	8.62	0.236	1.40	28.12
40MD + 60B	8.80	0.209	1.45	25.12
25S + 75B	8.56	0.251	1.31	44.10
25MD + 75B	8.72	0.225	1.25	39.91
20S + 80B	8.45	0.257	1.25	45.10
20MD + 80B	8.67	0.240	1.23	43.76
15S + 85B	8.38	0.266	1.20	46.71
15MD + 85B	8.61	0.246	1.22	46.10
10S + 90B	8.24	0.273	1.10	56.03
10MD + 90B	8.50	0.253	1.07	54.52
5S + 95B	7.97	0.281	1.11	54.84
5MD + 95B	8.18	0.258	1.10	51.96
100B	7.97	0.281	1.12	45.76

Table 4 Maximum swelling percentage of S–B and MD–B mixes

Bentonite (%)	Marble dust	Sand
	Max. swell (%)	Max. swell (%)
10	0.47	0.52
20	2.53	3.12
40	6.66	7.26
60	18.66	20.26
75	27.20	28.75
80	37.89	38.47
85	40.97	42.51
90	45.24	47.55
95	48.25	49.85
100	53.22	53.22

4.3 Maximum Swell Percentage of S–B and MD–B Mixes Subjected to Electrolyte Concentration

It is interesting to note that swelling percentage of both mixes reduces drastically subjected to the increase in the concentration of NaCl solutions (Fig. 6a). Bentonite is more sensitive to pore solutions because of its high surface area [33]. The increment in the swelling percentage is due to an attack of alkaline solution (Fig. 6b) on the soil minerals and, is mostly due to the more formation of hydrous sodium carbonate (Na_2CO_3) and sodium bicarbonate ($NaHCO_3$) in the interlayer spacing [34].

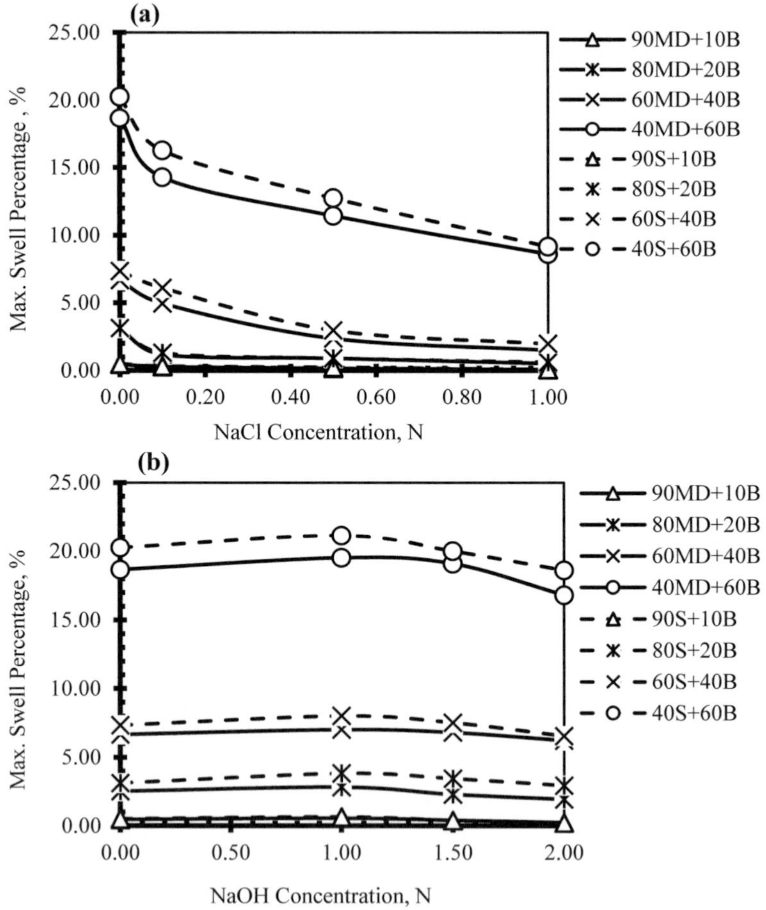

Fig. 6 Effect of electrolyte concentration **a** NaCl and **b** NaOH on maximum swelling presentation of S–B and MD–B mixes [35]

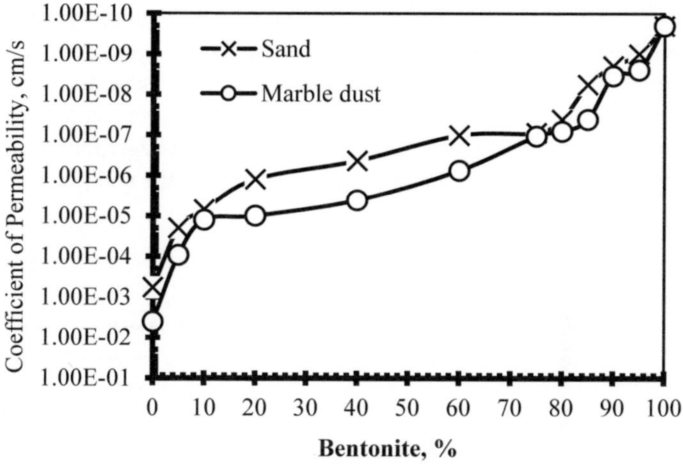

Fig. 7 Permeability of sand and marble dust with addition of bentonite content [25]

4.4 Permeability of S–B and MD–B Mixes Without Electrolyte Concentration

It is depicted from Fig. 7 that the permeability of sand and marble dust drastically decreases with the addition of bentonite content.

It is interesting to note that similar permeability behavior is observed for both S–B and MD–B mixes. Permeability of S–B mixes varies between 5.81×10^{-4} and 1.00×10^{-9} cm/s; while for MD–B mixes, it varies from 4×10^{-3} to 2.051×10^{-9} cm/s with increase in bentonite content up to 95%. However, other factors such as type of bentonite, bentonite content, void ratio, degree of saturation, swelling behavior are also reported to be key parameters in controlling the permeability of S–B mixes [36].

4.5 Permeability of S–B and MD–B Mixes Subjected to Electrolyte Concentration

The permeability behaviors of S–B and MD–B mixes subjected to varying concentrations of NaCl and NaOH solutions are presented in Fig. 8a, b, respectively. The ranges of coefficient of permeability values of uncontaminated S–B and MD–B mixes are pronounced to be 7.01×10^{-6} to 1.01×10^{-7} cm/s and 1.25×10^{-5} to 7.44×10^{-7} cm/s, respectively. It is observed from Fig. 8a that the permeability of both mixes enhances with an increase in the concentration of NaCl salt solutions. It has been reported that when the sample mixes come into the contact with the NaCl salt solutions, the DDL reduces, leading to the formation of cracks and thereby increase in the inter-granular pores [37–39]. It is observed from Fig. 8b that the permeability of both mixes reduces continuously with NaOH solutions up to 2N concentration. Nyamangara et al. [40] reported that the reduction in the permeability subjected to NaOH solutions is due to (i) decrease in the soil organic matter (SOM) continuously with the NaOH concentration and thereby diminishing the aggregate stability of the mixtures and (ii) an increase in the net negative charge of the soil particles with rise in pH.

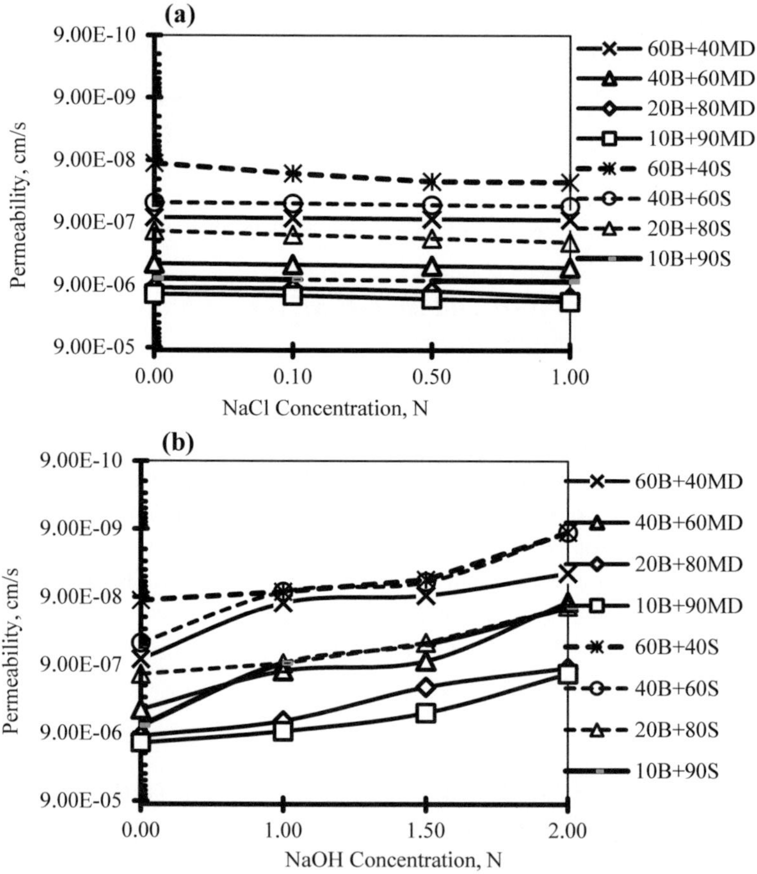

Fig. 8 Effect of electrolyte concentration **a** NaCl and **b** NaOH on permeability of S–B and MD–B mixes [35]

4.6 Microstructural and Elemental Examination of S–B and MD–B Mixes with and Without Electrolyte Concentration

Figure 9a–t show the microstructural examination of S–B and MD–B mixtures. Figure 9a–d illustrate the effect of bentonite content on the matrix of sand and marble dust. It is observed clearly that the accumulation of bentonite particles has been pronounced in the inter-particle and intraparticle voids of sand and marble dust. The microstructural images with 10% bentonite content illustrate that the filling and coating of bentonite particles within the inter-particle voids and the surface of the sand and marble dust particles have transformed the coarse-grained matrix to coarse-grained cohesive matrix (Fig. 9a–d). The microstructural examination validates the results obtained through experimental investigation on permeability and swelling for both S–B and MD–B mixes.

The SEM images of the S–B and MD–B mixes subjected to NaOH solutions have pronounced the disintegration of initial matrix as compared to uncontaminated

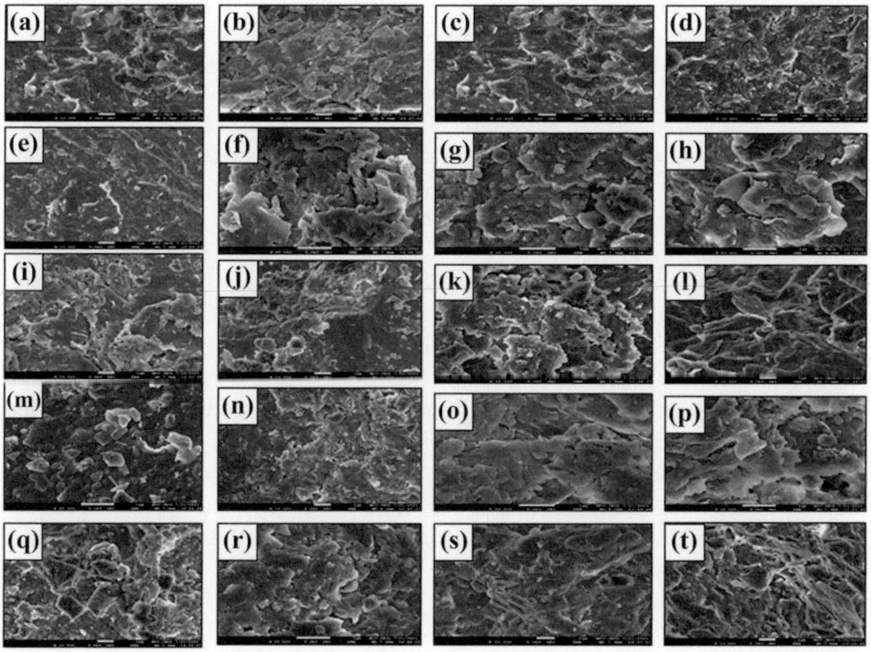

Fig. 9 SEM images of S–B and MD–B mixes without electrolyte concentration **a** 90MD + 10B, **b** 40MD + 60B, **c** 90S + 10B, **d** 40S + 60B, 1N NaOH **e** 90MD + 10B, **f** 40MD + 60B, **g** 90S + 10B, **h** 40S + 60B, 2N NaOH, **i** 90MD + 10B, **j** 40MD + 60B, **k** 90S + 10B, **l** 40S + 60B mixes, 0.1 N NaCl **e** 90MD + 10B, **f** 40MD + 60B, **g** 90S + 10B, **h** 40S + 60B, 1N NaCl, **i** 90MD + 10B, **j** 40MD + 60B, **k** 90S + 10B, **l** 40S + 60B mixes

samples with 1N NaOH (Fig. 9e–h). This is observed more in samples subjected to a higher concentration of 2N NaOH solution (Fig. 9i–l).

The NaCl-contaminated leachate destroys the uniform layer of smectite particles and initial matrix of samples. It is interesting to note the smaller size of voids in the microstructural images of both mixes exposed to 0.1N NaCl concentrations (Fig. 9m–p). Further, the voids are pronounced to be increased in the samples subjected to higher concentration of NaCl solution (Fig. 9q–t). Salt solution as a pore fluid increases the repulsion between the particles, resulting in an increase in the voids of the mixes and thereby corresponding increase in the permeability of samples.

5 Conclusion

Following are the conclusion drawn from the present study.

- The subgrade strength of BCS has been improved drastically with addition of MD up to 60%. Hence, 60% MD can be considered as an optimum content to improve the subgrade strength and CBR of BCS as per Indian Road Congress (IRC) for selection of subgrade materials.
- The pH of sand and MD increases up to 20% bentonite content and reduces thereafter. However, the electrical conductivity of sand and MD increases continuously with an increase in bentonite content. The pH of S–B and MD–B mixes decreases and the EC increases continuously with the concentrations of NaCl solution, respectively. On contrary, the S–B and MD–B mixes subjected to NaOH solutions increase the pH with reduction in the EC of the mixes, respectively.
- The maximum swell percentage of sand and marble dust has increased drastically by following a similar trend with varying bentonite contents. The maximum swell percentage of the S–B and MD–B mixes subjected to NaCl solutions reduces drastically. Further, reduction in swelling percentage of both mixes is pronounced only at higher NaOH concentration.
- The NaCl contamination of the S–B and MD–B mixes increases the coefficient of permeability. On contrary, coefficient of permeability of both mixes reduces continuously upon NaOH contamination. However, MD and sand need to be incorporated with at least 40% and 60% B to encounter the effect of electrolyte concentrations for maintaining the coefficient of permeability ($< 10^{-7}$ cm/s) in landfill liner, respectively.
- Microstructural examination validates that the variations in the swelling and permeability are due to the filling of inter-particle voids and adherence of bentonite particles on the surface of sand and marble dust. Microstructural examination also confirms the formation of voids in the samples with increasing the NaCl concentration, whereas the filling of the pores/voids are pronounced with NaOH contamination.

- Overall, the present study has brought out clearly the bulk and effective utilization of marble dust in geotechnical construction projects i.e., subgrade preparation, and landfill liner material. It is believed that utilization of marble dust, a waste material, also reduces the construction cost and saves the depletion of natural resources like sand as well as fruitful for embracing the sustainable development and safe ecosystem for environmentally friendly construction operations.

References

1. Kumar A, Walia BS, Bajaj A (2007) Influence of fly ash, lime, and polyester fibers on compaction and strength properties of expansive soil. J Mater Civ Eng 19(3):242–248
2. Soosan TG, Sridharan A, Jose BT, Abraham BM (2005) Utilization of quarry dust to improve the geotechnical properties of soils in highway construction. Geotechn Test J 28(4):391–400
3. Tozsin G, Oztas T, Arol AI, Kalkan E, Duyar O (2014) The effects of marble wastes on soil properties and hazelnut yield. J Clean Prod 81:146–149
4. Okagbue CO, Onyeobi TUS (1999) Potential of marble dust to stabilise red tropical soils for road construction. Eng Geol 53(3–4):371–380
5. Steinberg M (2000) Expansive soils and the geomembrane remedy. In: Advances in unsaturated geotechnics, pp 456–466
6. Nwonu DC, Ikeagwuani CC (2021) Evaluating the effect of agro-based admixture on lime-treated expansive soil for subgrade material. Int J Pavement Eng 22(12):1541–1555
7. Abdel-Shafy HI, Mansour MSM (2018) Solid waste issue: Sources, composition, disposal, recycling, and valorization. Egyp J Petrol 27(4):1275–1290
8. Bureau of Indian standards IS 2720 (Part 4) (1985) Methods of test for soils: grain size analysis, New Delhi, India
9. Bureau of Indian Standards IS 2720, (Part 3) (1980) Methods of test for soils: determination of specific gravity of soil, New Delhi, India
10. Bureau of Indian Standards (second revision) IS 2720, (Part 5) (1985) Methods of test for soils: determination of liquid limit and plastic limit, New Delhi, India
11. Bureau of Indian Standards (first revision) IS 2720 (Part 6) (1972) Methods of test for soils: determination of shrinkage factors, New Delhi, India
12. Sridharan A, Sivapullaiah PV (2005) Mini compaction test apparatus for fine grained soils. ASTM Geotech Test J 28(3):7
13. Bureau of Indian Standards IS 2720 (Part 16) (1979) Methods of test for soils: laboratory determination of CBR. New Delhi, India
14. Bureau of Indian standards (first revision) IS: 2720 (Part 10) (1973) Methods of test for soils: determination of unconfined compressive strength, New Delhi, India
15. Bureau of Indian Standards IS 2720, (Part 40) (1977) Methods of test for soils: determination of free swell index of soil, New Delhi, India
16. Sivapullaiah PV, Sitharam TG, Subba Rao KS (1987) Modified free swell index for clays. Geotech Test J 10(2):80–85
17. Bureau of Indian Standards IS 2720 (Part 26) (1987) Methods of test for soils: determination of pH of soil. New Delhi India
18. Bureau of Indian Standards IS 2720 (Part 13) (1986) Reaffirmed in 2016, Methods of test for soils: determination of shear strength. New Delhi India
19. Bureau of Indian Standards IS 2720, (Part XV) (1986) Methods of test for soils: determination of consolidation properties. New Delhi, India
20. Bureau of Indian Standards IS 2720, (Part 17) (1986) Methods of test for soils: determination of permeability of soil. New Delhi, India

21. Jain AK, Jha AK (2020) Improvement in subgrade soils with marble dust for highway construction: a comparative study. Indian Geotech J 50(2):307–317
22. Jha AK, Jain AK (2019) Potential of marble dust to improve the physical behavior of soil. In: Geotechnics for transportation infrastructure. Springer, Singapore, pp 189–201
23. Jain AK, Gupta M, Jha AK (2021) Soil amendment using marble waste for road construction. In: Problematic soils and geoenvironmental concerns. Springer, Singapore, pp 245–256
24. Jain AK, Jha AK (2020) Geotechnical behaviour and micro-analyses of expansive soil amended with marble dust. Soils Found 60(4):737–751
25. Jain AK, Jha AK, Akhtar PM (2022) Assessing the swelling and permeability behavior of novel marble dust–bentonite with sand–bentonite mixes for use as a landfill liner material. Indian Geotech J 52(3):675–690
26. Diouri K, Chaqroune A, Kherbeche A, Bentama J, Lahrichi A (2015) Kinetics of yellow dye adsorption onto marble powder sorbents. J Mater Environ Sci 6(1):79–92
27. Kumararaja P, Manjaiah KM, Datta SC, Binoy S (2017) Remediation of metal contaminated soil by aluminium pillared bentonite: synthesis, characterisation, equilibrium study and plant growth experiment. Appl Clay Sci 137:115–122
28. Van der Marel HW, Beutelspacher H (1976) Atlas of infrared spectroscopy of clay minerals and their admixtures. Elsevier Publishing Company
29. Janik LJ, Merry RH, Forrester ST, Lanyon DM, Rawson A (2007) Rapid prediction of soil water retention using mid infrared spectroscopy. Soil Sci Soc Am J 71(2):507–514
30. Bureau of Indian Road Congress IRC (2012) 37: Tentative guidelines for the design of flexible pavements. New Delhi India
31. Bureau of Indian Road Congress IRC (2018) 37: Guidelines for the design of flexible pavements (Fourth rev.) New Delhi India
32. Bureau of Indian Road Congress IRC (2007) 72: Guidelines for the design of low volume rural roads. New Delhi India
33. Chalermyanont T, Surapon A, Nantanit C (2007) Transport parameters and heavy metal adsorption of a compacted lateritic soil. Eng Appl Sci Res 34(1):73–87
34. Kuriakose LP, Madheeha A, Francis A, Baby BJ, Mowshmi S (2018) Comparative study on performances of clay-bentonite and bentonite-quarry dust mixture as landfill liner. Int J Eng Res Appl 8(3):2248–9622
35. Jha AK, Jain AK, Akhtar MP (2022) Evaluating the behavior of marble dust-bentonite as an alternative to sand–bentonite mixes permeated with electrolyte concentrations. J Hazar Toxic Radioactive Waste 27(1):04022035
36. Jha AK, Sivapullaiah PV (2015) Susceptibility of strength development by lime in gypsiferous soil—a micro mechanistic study. Appl Clay Sci 115:39–50
37. Karimi E, Abdolzadeh A, Sadeghipour HR, Aminei A (2012) The potential of glauconitic sandstone as a potassium fertilizer for olive plants. Arch Agron Soil Sci 58(9):983–993
38. Hartman M, Svoboda K, Pohořelý M, Šyc M (2013) Thermal decomposition of sodium hydrogen carbonate and textural features of its calcines. Industr Eng Chem Res 52(31):10619–10626
39. Gleason MH, Daniel DE, Gerald RE (1997) Calcium and sodium bentonite for hydraulic containment applications. J Geotechn Geoenviron Eng 123(5):438–445
40. Young JH, Katsumi T, Benson CH, Edil TB (2001) Hydraulic conductivity and swelling of nonprehydrated GCLs permeated with single-species salt solutions. J Geotech Geoenviron Eng 127(7):557–567

Finite Element Analysis of Slope Stability Considering Soil Strength Anisotropy

Zhi Li, Kefeng Xiao, and Yi He

Abstract In order to analyze the influence of soil anisotropy on slope stability analysis, an anisotropy yield criterion is introduced to evaluate the influence of varying shear strength on slope stability. The anisotropic characteristics of soil can be represented by the parameters n and β, which can be determined through experimental measurements. This paper constructs a complete finite element analysis framework for slope stability to accurately obtain the safety factor of slope (F_s). It is found that neglecting the anisotropy of soil will overestimate the stability of slope, especially when the slope angle (α) is small. The smaller the value of anisotropic parameter n is, the smaller F_s is. F_s will first decrease with the increase of anisotropic parameter β. When β increases to a critical value, F_s will increase with the increase of β. The greater the α is, the greater the critical value is.

Keywords Slope stability · Soil anisotropy · Finite element analysis · Safety factor · Anisotropic yield criterion

1 Introduction

The failure of the slope will pose a great threat to the safety of people's lives and property. At present, the technology and method of obtaining the safety factor of the isotropic slope have been quite mature [1–7]. However, some studies have found that neglecting the anisotropy of soil can lead to overestimations of slope stability. How to accurately analyze the influence of soil anisotropy on slope stability analysis is one of the hot research topics nowadays.

Casagrande and Carillo [8] first defined the anisotropy, which means that the strength and deformation characteristics of soil at the same position were different along different loading directions. Under the influence of anisotropy, the soil strength

Z. Li · K. Xiao (✉) · Y. He
Faculty of Geosciences and Environmental Engineering, Southwest Jiaotong University, Chengdu, Sichuan, China
e-mail: xkfswjtu@163.com

H. Hazarika et al. (eds.), *Multidisciplinary Approach for Disaster Resilience and Sustainable Technologies*, Lecture Notes in Civil Engineering 642,
https://doi.org/10.1007/978-981-96-8618-6_15

at different positions of the slope is obviously different. Koutsoftas and Ladd [9] suggested using the shear strength at the top, the middle, and the bottom of the slope to evaluate the stability of the slope. Some scholars [10, 11] found that the influence of strength anisotropy on gentle slope was more obvious. Neglecting soil anisotropy would lead to a larger safety factor [12, 13]. In the early stage, the triaxial test was mostly used to explore the strength anisotropy of soil [14, 15], but Saada [16] pointed out that the conventional test method could not achieve the real rotation of principal stress axis, and recommended the hollow cylinder torsional shear test. However, most of the above-mentioned studies primarily focus on the anisotropy of the undrained shear strength of the soil. According to the results of the hollow cylinder torsional shear test, Yuan et al. [17] proposed a new yield criterion considering the anisotropy of friction Angle, which is relatively simple, and the anisotropy parameters can be easily measured through experiments.

Therefore, based on the finite element method, this paper introduces the anisotropy yield criterion [17] to explore the influence of friction angle anisotropy on slope stability, and discusses the influence of soil anisotropy parameters and slope angle.

2 The Anisotropic Yield Criterion

The strength anisotropies of soils are usually classified into two types, the inherent and induced ones. A new anisotropic yield criterion proposed by Yuan et al. [17] is introduced to discuss the influence of primary anisotropy of soil on slope stability in this paper. The anisotropic yield surface is shown in Fig. 1.

The anisotropic yield function can be explicitly given by

$$f\left(\sigma_x, \sigma_y, \sigma_{xy}\right) = R + F(p, \Theta) \tag{1}$$

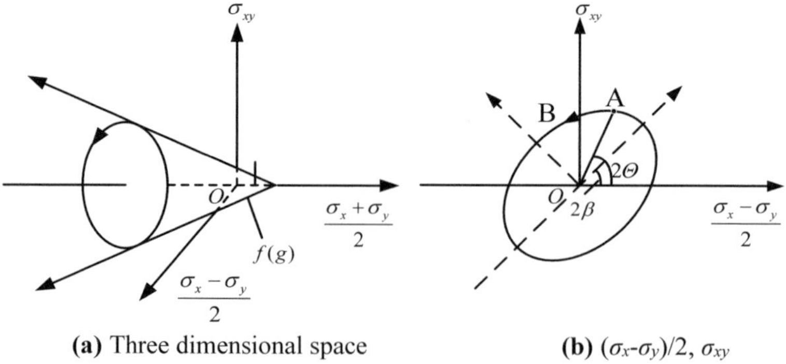

Fig. 1 Yield surface in the stress space

$$R = \sqrt{\left(\frac{\sigma_x - \sigma_y}{2}\right)^2 + \sigma_{xy}^2} \tag{2}$$

$$F(p, \Theta) = (p - c\cot\varphi_{\max})\sin\varphi(\Theta) \tag{3}$$

The internal friction angle of soil under different principal stress directions is different, and the calculation formula is

$$\sin\varphi(\Theta) = \frac{n\sin\varphi_{\max}}{\sqrt{n^2\cos^2(2\Theta - 2\beta) + \sin^2(2\Theta - 2\beta)}} \tag{4}$$

$$p = \left(\sigma_x + \sigma_y\right)/2 \tag{5}$$

$$\tan(2\Theta) = 2\sigma_{xy} \big/ \left(\sigma_x - \sigma_y\right) \tag{6}$$

where f is the anisotropic yield function, R is the distance between the stress point and the origin of coordinates, c is cohesion, $\varphi_{\max}$ is maximum peak internal friction angles, Θ is the direction angle of principal stress. β and n are the anisotropic parameters, which can be measured by the hollow cylinder torsional shear tests. Further, β is the direction angle corresponding to the maximum internal friction angle. β ranges from 0 to $\pi/4$. n is the ratio of the sine of the maximum friction angle to the minimum friction Angle, $n = (\sin\varphi_{\min})/(\sin\varphi_{\max})$, which ranges from 0 to 1. The smaller n is, the more obvious the anisotropy is. When $n = 1$, $\sin\varphi(\Theta) = \sin\varphi_{\max}$, f becomes the classical isotropic Mohr–Coulomb yield function. The effect of anisotropic parameters on the internal friction angle is shown in Fig. 2.

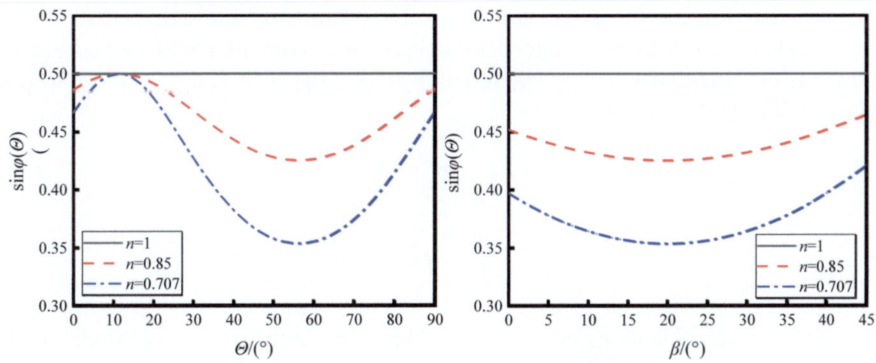

(a) $\varphi_{\max}$=30°, β=11.25° **(b)** $\varphi_{\max}$=30°, Θ=65°

Fig. 2 The effect of anisotropic parameters on internal friction angle

3 Finite Element Analysis

Finite element method is a very effective method to obtain the safety factor of slope. The solution of this kind of elastic–plastic problem is equivalent to the solution of a large nonlinear equation group, which can be solved by iteration. In this paper, the elasto-plastic problem is solved by the method of constructing body load [18]. Eight-node quadrilateral element is selected as the basic element to construct the finite element solution model. The stiffness equation

$$Ku = P_a \tag{7}$$

is established to solve the displacement increment. Then the stress increment can be solved according to the displacement increment.

$$\Delta\sigma = D \cdot \left(Bu_i - \delta\varepsilon^{vp}\right) \tag{8}$$

where B is strain–displacement matrix, D is elastic stress–strain matrix, $\delta\varepsilon^{vp}$ is plastic strain which is zero initially. Substituting this stress increment into the yield function, if $f \geq 0$, plastic strain will occur in the soil.

$$\delta\varepsilon^{vp} = \Delta t \cdot f \cdot \frac{\partial Q}{\partial \sigma} \tag{9}$$

$$\Delta t = \frac{4(1+\nu)(1-2\nu)}{E\left(1-2\nu+\sin^2\varphi\right)} \tag{10}$$

Q is the plastic potential function, which is considered to be equal to the yield function due to using the associated flow rule. When solving the derivative of the plastic potential function a rounded hyperbolic approximation was carried out to eliminate the singularity of the yield function at the edge or tip [2]. Then the stiffness equation will be replaced by

$$Ku = P_a + P_b \tag{11}$$

$$P_b = \int B^T D^e \delta\varepsilon^{vp} \mathrm{d}\Omega \tag{12}$$

where P_a is external load, P_b is body loads. The calculation flow chart of slope safety factor is shown in Fig. 3. This paper employs a binary search method to determine the safety factor of the slope. FS_{max} and FS_{min} in the figure represent the upper and lower bounds of the reduction factors used in the binary search. They continuously change through iterations until a safety factor meeting the allowable error is found.

One numerical example of typical soil slope models is used to analyze the effects of strength anisotropy on slope stability [19]. The eight-node quadrilateral element is

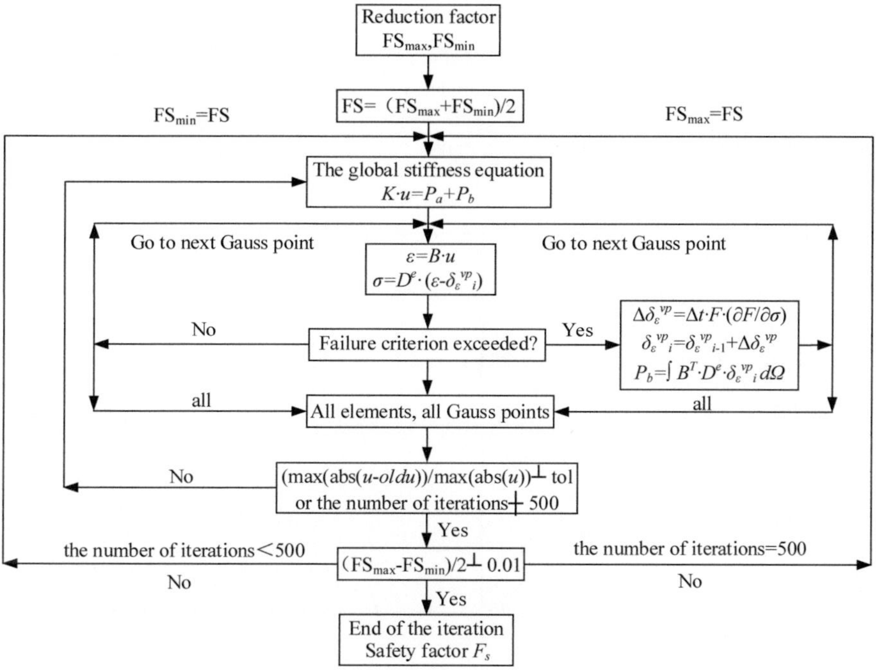

Fig. 3 The calculation flow chart of slope safety factor

utilized to discretize the slope model, with a total number of 1000 elements (Fig. 4). The properties of the slope are listed in Table 1.

The anisotropic parameters followed Yuan et al. [17]. Considering soil anisotropy, the safety factor of slope is obtained (Fig. 5).

Considering the soil as isotropic ($n = 1$), safety factor (F_s) is 1.255. The calculation result of the existing numerical software OPTUM G2 is 1.257, which proves that the method of obtaining the slope safety factor in this paper is effective. It can be found

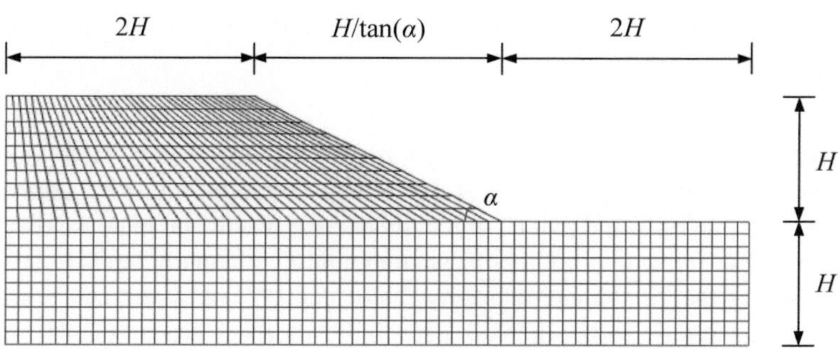

Fig. 4 Finite element model of slope

Table 1 The properties of the slope

Unit weight γ (kN/m^3)	Friction angle φ_{max} (°)	Cohesion c (kPa)	Young's modulus E (kN/m^2)	Poisson's ratio Y	Hight H (m)	Slope angle α (°)
20	10	10	10,000	0.3	5	26.565

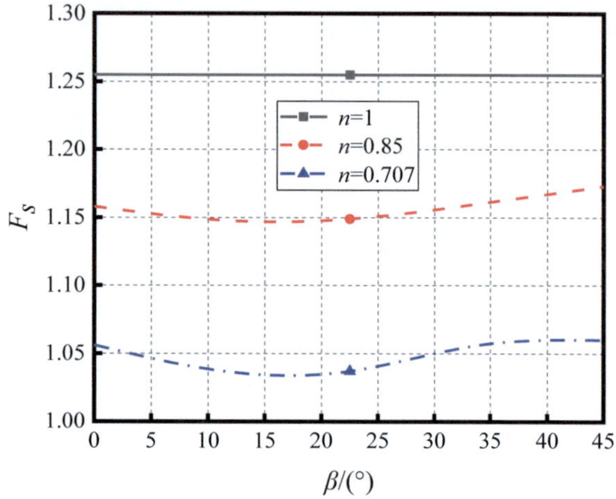

Fig. 5 The effect of anisotropic parameters on slope safety factor

from Fig. 5 that when the anisotropy of soil is considered, the F_s is smaller than that obtained under the isotropic condition. F_s first decreases and then increases with the increase of anisotropy parameter β. F_s decreases with the increase of anisotropy parameter n. The value of n has a great influence on the safety factor of slope.

The plastic strain nephograms of the slope for isotropic and anisotropic conditions are displayed in Fig. 6. When the plastic zone extends from the bottom of the slope to the top of the slope, it can be considered that the slope is in the ultimate failure state [20]. The smaller n is, the smaller F_s is.

4 Parameters Discussion

The effect of anisotropic and geometric parameters on slope safety factor is further discussed in this paper. Figure 7 shows the influence of slope angle (α) on the slope safety factor. It can be found that F_s gradually decreases with the increase of α, when α is larger, the decreasing amplitude is smaller, and F_s gradually tends to be stable. Figure 8 shows the percentage of safety factor gap (P_s) between $n = 1$ and $n = 0.707$. P_s will decrease with α increase especially when α is greater than 50°, which indicates that the influence of anisotropy on the F_s is smaller when α is smaller.

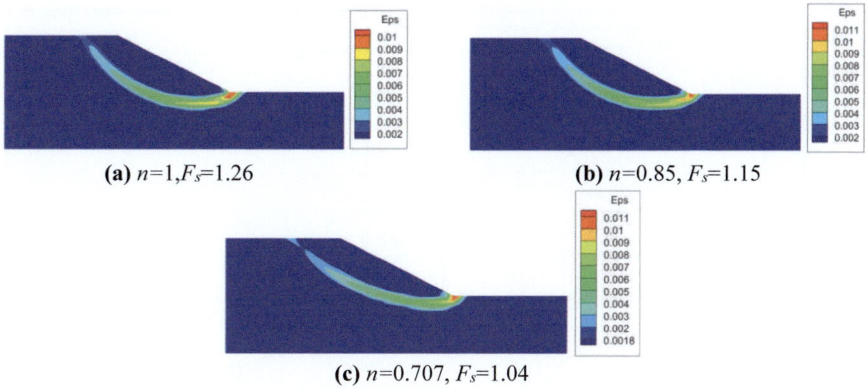

Fig. 6 The plastic strain nephograms of the slope ($\beta = 11.25°$)

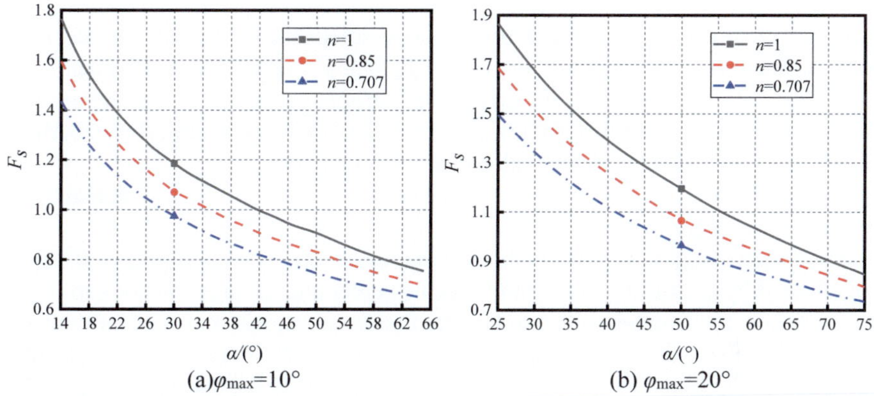

Fig. 7 The effect of slope angle on safety factor ($\beta = 11.25°$)

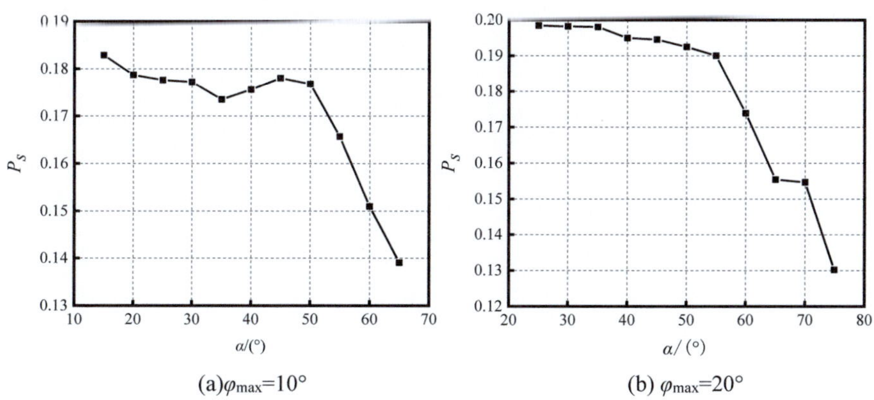

Fig. 8 The percentage of safety factor gap between $n = 1$ and $n = 0.707$ ($\beta = 11.25°$)

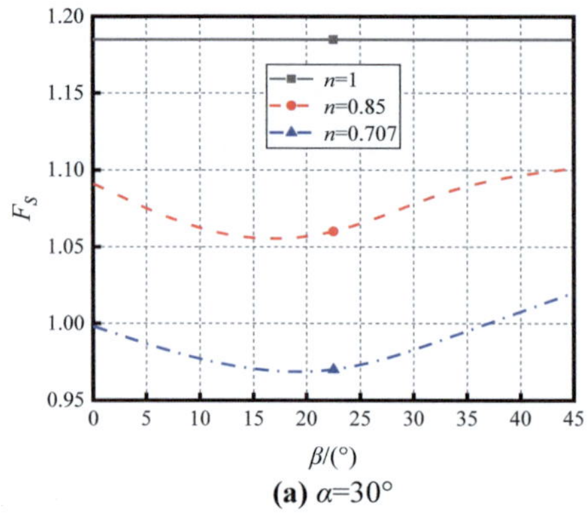

Fig. 9 The effect of slope angle on safety factor

Figure 9 shows the effect of β on the slope safety factor. F_s will first decrease with the increase of anisotropic parameter β. When β increases to a critical value (β_0), F_s will increase with the increase of anisotropic parameter β. The greater α is, the greater β_0 is. For instance, When $\alpha = 30°$, $\beta_0 = 17.11°$; $\alpha = 45°$, $\beta_0 = 26.10°$; $\alpha = 60°$, $\beta_0 = 34.74°$. The critical value is almost unaffected by n.

Figure 10 shows the effect of φ_{max} on the slope safety factor. F_s will increase with the increase of φ_{max}.

The greater φ_{max} is, the more obvious the anisotropy is. When $\varphi_{max} = 0°$, F_s is not affected by anisotropy. It is found that the percentage of safety factor gap (P_s) is greater when α is smaller, which indicating that anisotropy is more obvious when α is small. For instance, When $\varphi_{max} = 20°$ and $\alpha = 30°$, $P_s = 19.8\%$ ($P_s = (1.865 - 1.495)/1.865 = 19.8\%$) between $n = 1$ and $n = 0.707$. However, when $\varphi_{max} = 20°$ and $\alpha = 60°$, $P_s = 13.0\%$ ($P_s = (0.845 - 0.735)/0.845 = 13.0\%$).

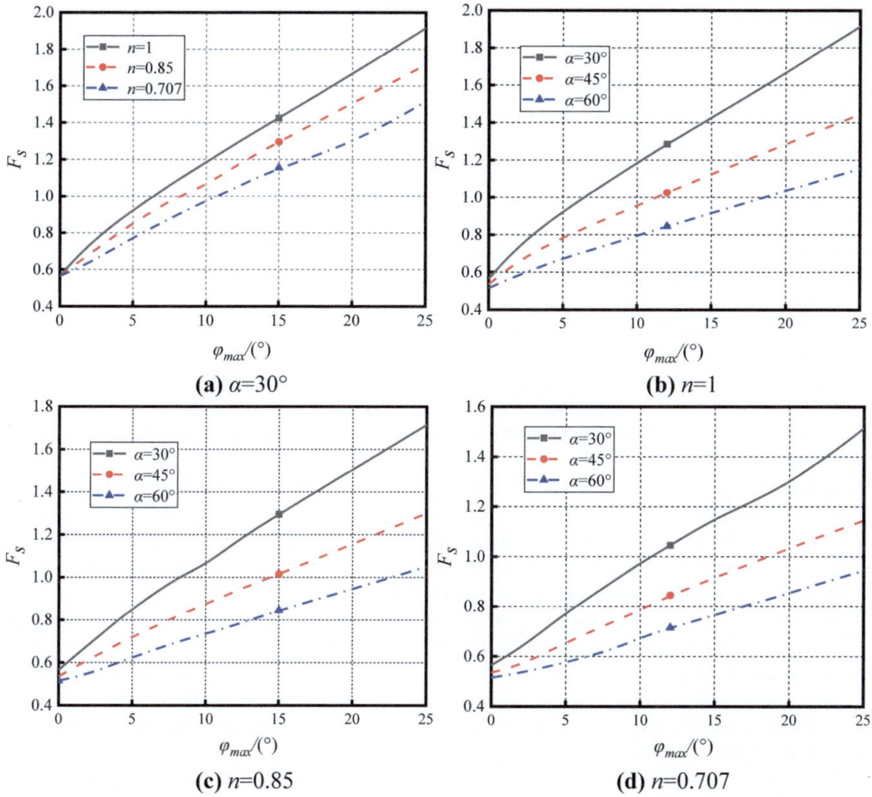

Fig. 10 The effect of the maximum internal friction angle on safety factor ($\beta = 11.25°$)

5 Conclusion

Based on the finite element method, the anisotropic yield criterion is introduced to analyze the influence of soil anisotropy on slope stability. It is found that neglecting the anisotropy of soil will overestimate the stability of slope, especially when the slope angle (α) is small. The smaller the value of anisotropic parameter n is, the smaller the safety factor of slope (F_s) is. F_s will first decrease with the increase of anisotropic parameter β. When β increases to a critical value, F_s will increase with the increase of anisotropic parameter β. The greater α is, the greater the critical value is.

Acknowledgements This study was supported by the National Natural Science Foundation of China (Grant Nos. 42177128, 42077236, 42011530170, 52278413), and by the Sichuan Science and Technology Program (Grant Nos. 2021YFH0037, 2021YFH0320, 2019JDJQ0018).

References

1. Sloan SW (1987) Substepping schemes for the numerical integration of elastoplastic stress-strain relayions. Int J Numer Meth Eng 24:893–911
2. Abbo AJ (1997) Finite element algorithms for elastoplasticity and consolidatio. University of Newcastle, Australia
3. Sloan SW (2013) Geotechnical stability analysis. Géotechnique 63(7):531–572
4. He Y, Hazarika H, Yasufuku N, Zheng Han LYG (2015) Three-dimensional limit analysis of seismic displacement of slope reinforced with piles. Soil Dyn Earthq Eng 77(10):446–452
5. He Y, Liu Y, Zhang YB, Yuan R (2019) Stability assessment of three-dimensional slopes with cracks. Eng Geol 252:136–144
6. He Y, Liu Y, Hazarika H, Yuan R (2019) Stability analysis of seismic slopes with tensile strength cut-off. Comput Geotech 112:245–256
7. He Y, Yu JY, Yuan R et al (2022) Stability and failure mechanisms in three-dimensional cracked slope: static and dynamic analysis. Comput Geotechn 144
8. Casagrande A, Carillo N (1944) Shear failure of anisotropic materials. J Boston Soc Civ Eng 31(04):74–87
9. Koutsoftas DC, Ladd CC (1985) Design strengths for an offshore clay. J Geotechn Eng 111(03):337–355
10. Lo KY (1965) Stability of Slopes in anisotropic soils. Atmos Chem Phys 12(23):11679–11694
11. Hwang J, Dewoolkar M, Ko HY (2002) Stability analysis of two-dimensional excavated slopes considering strength anisotropy. Can Geotech J 39(5):1026–1038
12. Shogaki T, Kumagai N (2008) A slope stability analysis considering undrained strength anisotropy of natural clay deposits. Soils Found 48(06):805–819
13. Aghajani HF, Salehzadeh H, Shahnazari H (2015) Stability analysis of sandy slope considering anisotropy effect in friction angle. Sadhana-Acad Proc Eng Sci 40(6):1955–1974
14. Hansen JB, Gibson RE (1949) Undrained shear strengths of anisotropically consolidated clays. Geotechnique 1(3):189–200
15. Bjerrum L, Eide O (1956) Stability of strutted excavations in clay. Geotechnique 6(1):32–47
16. Saada AS (1970) Testing of anisotropic clay soils. J Soil Mech Found Div 96(5):1847–1852
17. Yuan R, Yu HS, Yang DS, Hu N (2019) On a fabric evolution law incorporating the effects of b-value. Comput Geotech 105:142–154
18. Smith IM, Griffith DV, Margetts L (2015) Programming the finite element method. John Wiley and Sons
19. Fenton GA, Griffiths DV (2008) Risk assessment in geotechnical engineering. John Wiley and Sons, Hoboken
20. He Y, Wang XY, Yuan R, Liu KW, Zhuang PZ (2019) On the computational precision of finite element algorithms in slope stability problems. Math Probl Eng 4:1–15

MIX
Papier aus verantwortungsvollen Quellen
Paper from responsible sources
FSC® C105338

If you have any concerns about our products,
you can contact us on
ProductSafety@springernature.com

In case Publisher is established outside the EU,
the EU authorized representative is:
Springer Nature Customer Service Center GmbH
Europaplatz 3, 69115 Heidelberg, Germany

Printed by Libri Plureos GmbH
in Hamburg, Germany